Unmanned Aircraft Systems for Logistics Applications

John E. Peters, Somi Seong, Aimee Bower, Harun Dogo, Aaron L. Martin, Christopher G. Pernin

Prepared for the United States Army

RAND ARROYO CENTER

The research described in this report was sponsored by the United States Army under Contract No. W74V8H-06-C-0001.

Library of Congress Cataloging-in-Publication Data

Unmanned aircraft systems for logistics applications / John E. Peters ... [et al.].
    p. cm.
  Includes bibliographical references.
  ISBN 978-0-8330-5044-1 (pbk. : alk. paper)
  1. Drone aircraft. 2. United States. Army—Equipment and supplies.
  3. Logistics. I. Peters, John E., 1947-

    UG1242.D7U5655 2012
    358.4'483—dc23

                                                              2011047888

The RAND Corporation is a nonprofit research organization providing objective analysis and effective solutions that address the challenges facing the public and private sectors around the world. RAND's publications do not necessarily reflect the opinions of its research clients and sponsors.

RAND® is a registered trademark.

Published 2011 by the RAND Corporation
1776 Main Street, P.O. Box 2138, Santa Monica, CA 90407-2138
1200 South Hayes Street, Arlington, VA 22202-5050
4570 Fifth Avenue, Suite 600, Pittsburgh, PA 15213-2665
RAND URL: http://www.rand.org/
To order RAND documents or to obtain additional information, contact
Distribution Services: Telephone: (310) 451-7002;
Fax: (310) 451-6915; Email: order@rand.org

Preface

The concept of operations for unmanned aircraft systems (UAS)[1] published by the Deputy Secretary of Defense and the Vice Chairman of the Joint Chiefs of Staff is optimistic about future roles for UAS. These roles include intelligence, surveillance, and reconnaissance (ISR); reconnaissance, surveillance, and target acquisition (RSTA); laser designation; attack; damage assessment; chemical, biological, radiological, nuclear, and explosive (CBRNE) detection and monitoring; cargo delivery and logistics resupply; and communications gateway extension. The concept of operations also notes that UAS may have roles in psychological operations, combat identification, early warning, locating enemy military equipment, monitoring borders, detecting mines, and supporting law enforcement.[2] Similarly, the National Defense Authorization Act of 2008 noted the appropriation committee's supportiveness of "efforts to explore the use of unmanned systems in a variety of roles and missions on the battlefield."[3] It is, therefore, only natural to analyze their utility for logistics applications. The question is: Among all

---

[1] Formally, "unmanned aerial vehicle" or "UAV" refers to the aircraft alone. "UAS" refers to the entire system, including the ground station, launch-and-recovery system, and organic maintenance elements.

[2] Department of Defense, *Joint Concept of Operations for Unmanned Aircraft Systems*, Washington, D.C.: Joint Unmanned Aircraft Systems Center of Excellence, November 2008, p. II-7.

[3] See 110th Congress 1st Session, *National Defense Authorization Act for Fiscal Year 2008*, 35-737, Senate Report 110-77, June 5, 2007.

the tasks confronting today's Army logisticians, which ones might be accomplished better or at lower risk through the introduction of UAS?

This monograph describes research that examined potential uses of UAS in a wide range of Army logistics applications, other than resupply, which is being studied separately.[4] The research reported here examined the technical and operational feasibility of UAS for a broad range of logistics applications. For those found feasible, we determined the cost and benefits of the UAS concepts relative to other, non-UAS options for accomplishing the same tasks. We also identified factors and conditions that bear on the relative cost-effectiveness of UAS concepts.

The Deputy Chief of Staff, G-4, Headquarters Department of the Army G-4, and the Commanding General of the U.S. Army Combined Arms Support Command (CASCOM) sponsored the research, which was carried out in RAND Arroyo Center's Military Logistics Program. RAND Arroyo Center, part of the RAND Corporation, is a federally funded research and development center sponsored by the United States Army.

The Project Unique Identification Code (PUIC) for the project that produced this document is ATFCR09990.

For more information on RAND Arroyo Center, contact the Director of Operations (telephone 310-393-0411, extension 6419; fax 310-451-6952; e-mail Marcy_Agmon@rand.org), or visit Arroyo's website at http://www.rand.org/ard/.

---

[4] General Dynamics was tasked to study the feasibility, costs, and benefits of UAS for emergency and routine resupply of Army units.

Contents

Figures

Tables

Summary

This project evaluated ten potential logistics applications for UAS to determine whether they are technically feasible, operationally feasible, and more cost-effective than other options. This study concentrated on reconnaissance and surveillance tasks to secure logistics convoys through overwatch with UAS; route reconnaissance looking for hazards that might endanger logistics convoys; and surveillance of pipelines, electrical lines, rivers, supply depots, disaster scenes, and predeployment theater reconnaissance. (A separate study, conducted by General Dynamics,[5] evaluated the use of UAS for emergency and routine resupply tasks.) This study also examined the potential for using UAS to locate airdropped cargo that misses the drop zone and to retrograde critical unserviceable items expeditiously.

We used a combination of Army data, interviews, and prior research to compile information for analysis. The research team made use of Combined Information Data Network Exchange (CIDNE) and FusionNet databases to sample the frequency and intensity of enemy attacks on logistics convoys and assets. We drew on recent research at RAND for insights into installation and pipeline security, the improvised explosive device (IED) problem, and the optimal employment of UAS. We interviewed Army personnel recently returned from operational theaters with practical experience in logistics operations, and we

---

[5] General Dynamics, *AR 5-5 Study: Future Modular Force Resupply Mission for Unmanned Aircraft Systems (UAS)*, prepared for Commanding General, Combined Arms Support Command and Department of the Army, G-4, General Dynamics Information Technology, February 24, 2010.

interviewed civilian contractors tasked with the security of pipelines and electrical lines in Iraq to understand those tasks more fully. Based upon the available data and our best efforts to understand current threats, we made estimates to help us determine which UAS applications are likely feasible and infeasible, beneficial and not, and cost-effective.

The study looks out to the year 2024 but also discusses how technology changes beyond that could affect the value of UAS. For many reconnaissance and surveillance tasks, we found that with current technology and costs, UAS are either not feasible or not cost-effective options in conditions similar to those experienced in Iraq; however, the case improves with conditions more like those found in Afghanistan. In all environments, the relative value of UAS will improve with the miniaturization and improvement of sensors; the appearance of new, small, less-expensive UAS; and improved network bandwidth. The relative values of options are highly sensitive to the value that the United States places on damages from enemy attacks and to the cost structure of UAS-based solutions for detecting and avoiding enemy action. Thus, the Army should continue to evaluate its options through the lens of life-cycle costs for UAS relative to the cost avoidance they provide by detecting potentially deadly and destructive enemy activities.

Table S.1 summarizes general findings about the feasibility and benefits of UAS for logistics applications.

The table lists six considerations and their tendencies in terms of favoring or not favoring UAS-based solutions. The first is cost. As the systems become cheaper or sensors become smaller, allowing small UAS to be used for logistics applications, UAS-based solutions are more favorable. But if the systems remain expensive, UAS-based solutions are not favorable for the challenges confronting logistics units.

Terrain is the second factor. If future Army operations take place in theaters where complex terrain and long distances render other alternatives for reconnaissance and surveillance infeasible, UAS are a good solution. If future operations most frequently occur in theaters with more open terrain and shorter distances that lend themselves to non-UAS solutions, then UAS would be less attractive.

Enemy tactics, techniques, and procedures (TTPs) are the third consideration. Circumstances in which adversaries tend to mass, or

Table S.1
Factors Influencing UAS-Based Solutions

Consideration	Factors Favoring UAS-Based Solutions	Factors Not Favoring UAS-Based Solutions
Cost of UAS	Cheaper (e.g., class II) systems	Expensive (e.g., Predator derivative) systems
Terrain	Complex, line-of-sight issues; large scale that renders fixed cameras, etc., infeasible	Open, open-broken terrain; relatively small scale that enables non-UAS solutions
Enemy TTPs	Tendency to mass, linger near site of interest to them	Short-dwell operations, low mass, refusal to hold terrain
Value of damage or loss that could be avoided through reconnaissance and surveillance	High costs from enemy actions	Low costs resulting from enemy actions
Weather	Conditions conducive to UAS flight and sensor operations	Conditions that hinder flight or sensor operations
Bandwidth	Low bandwidth puts premium on point-to-point UAS-operator systems	Abundant bandwidth that supports robust networks

linger, or otherwise present relatively easily observed behaviors would favor the employment of UAS. If the enemy minimizes its profile by conducting hit-and-run, short-dwell attacks, refuses to mass, and refuses to hold terrain or to become clearly associated with a piece of terrain, those behaviors are unfavorable for UAS operations.

The value of damage or loss that could be avoided through reconnaissance and surveillance may be among the most sensitive of considerations bearing on the utility of UAS. The U.S. experience in recent operations in both Afghanistan and Iraq reflects large swings in the numbers of incidents and the numbers of IED events over the course of time. If a future enemy were to surge and sustain attacks inflicting high costs on U.S. forces, such circumstances favor the use of UAS. The opposite development would also hold true, and low costs from enemy action would not favor UAS-based solutions.

The fifth consideration is weather. Conditions that do not challenge aircraft flight parameters (e.g., wind velocity, wind shear, very cold temperatures) or sensor operating parameters (e.g., minimum safe altitudes, cloud cover, rain, lightning), tend to favor the use of UAS. If weather is severe enough to interfere with flight operations or sensor performance, those conditions are unfavorable for UAS-based reconnaissance and surveillance solutions.

The final consideration is bandwidth. If a future theater is austere and bandwidth is therefore scarce (and perhaps oversubscribed), then simple, point-to-point, UAS-to-operator systems could prove valuable, depending upon the other considerations treated in this discussion. On the other hand, if bandwidth is abundant and supports robust networks, then specific logistics-oriented UAS-based approaches would be less attractive, and the priority would be on ensuring access to the theater network and the information available from shared assets.

For example, the benefit of UAS will depend in part on how well other elements of the command, control, communications, computers, intelligence, surveillance, and reconnaissance (C4ISR) network perform in the future. For instance, if the Under Secretary of Defense for Intelligence (USD(I)) Lieutenant General John Koziol's 2024 vision of extremely high-bandwidth networked C4ISR eventuates and the Army can operate in Afghanistan and future theaters supported by a much richer, denser C4ISR network, then ownership of UAS for any specific tasks will be less critical because the network will quickly provide information from all of them in a theater.

The following summarizes our assessments of the potential for applying UAS to specific logistical missions:

- The likely value of using UAS for convoy overwatch—the practice of shadowing a convoy with an armed UAS as the convoy proceeds along its route—varies with theater conditions. In conditions similar to those experienced in Iraq, we found this application to be operationally and technically infeasible. In particular, the short enemy dwell times and long distances limit value. However, conditions similar to those in Afghanistan, where the length of convoy routes and the difficulty of the terrain render non-UAS solutions impractical (e.g., the United States could not maintain mast-mounted cameras for the circumference of the ring road), are more favorable for the use of UAS.

- The promise of using UAS for route surveillance also varies by theater conditions. Given enemy behavior in Iraq, UAS have a probability of detecting enemy presence along routes and pipelines of just roughly 11 percent and less near the end of the UAS's orbit, when it has to turn around, leading to low UAS value.[6] Afghanistan offers different challenges. There, enemy behavior, which enables longer dwell times near targets, is more favorable to the use of UAS. Less favorable in Afghanistan are the weather conditions and terrain (e.g., high winds, rain, and steep ridges that may break line-of-sight or cause other issues). Technology might provide some relief, for example, if the Army is able to deploy long-endurance aerial communications relays to overcome line-of-sight problems.

- Ascertaining river navigability using light detection and ranging (LIDAR) aboard a UAS is technically infeasible, although this assessment could change if LIDAR systems can be successfully miniaturized and ruggedized to maintain calibration.

- UAS have a cost-effective role in fixed-site security as an integral part of larger security and surveillance systems typically dominated by radars and fixed cameras.

- UAS could be flown in support of theater reconnaissance prior to the deployment of Army forces. Doing so raises questions of operations security (OPSEC) because the UAS's presence, if detected, might lead the enemy to conclude that U.S. military action is imminent. UAS do not appear to offer an advantage, though, over the many Defense Intelligence Agency (DIA) and National Ground Intelligence Center (NGIC) operations support products that would be available to answer logisticians' questions about a new theater.

- UAS could be valuable for locating cargo that misses a drop zone.

- Retrograde of critical unserviceable items via UAS is feasible but does not deliver a clear benefit for most items. This is because reducing the time of evacuation to maintenance does not reduce

---

[6] There is, however, utility in discovering pipeline damage early, and a potential deterrent effect from having UAS observed operating in the area.

the number of spares the Army must buy or reduce the amount of uncertainty about the availability of spares, key sources of cost in the supply chain. The extreme vast majority of the retrograde time for sustainment maintenance to repair reparable spares arises from the supporting brigade supply support activity back to sustainment maintenance, not the first leg on the battlefield.

After careful examination of common themes found in the UAS concepts, we have concluded that the fundamental near-term force protection problem confronting the logistics community is situational awareness: the ability to understand one's environment, to detect threats, and to know one's own location and the locations of the enemy and friendly units. Situational awareness has become key to survival for U.S. Army units, including logistics formations. Logistics units no longer operate in secure rear areas where enemy threats are minimal; instead they face the same threats as combat formations once they leave the safety of a forward operating base's (FOB's) perimeter. Logistics units do not, however, have the same density of command, control, communications, computers, and intelligence (C4I) equipment as combat arms units and, as a result, have more difficulty sustaining robust situational awareness. It is this situational awareness deficit that some have proposed to overcome with UAS, which do have a role to play in a network that includes other assets.

Acknowledgments

Many people provided help with this research. We are grateful to those in the aviation and ISR industries who met with us to explain their evolving concepts and future products. We also benefited from a professional conference sponsored by the American Helicopter Society at which we were able to interact with a wide variety of academic and industry leaders involved in rotary-winged flight. We are indebted to Dr. Robert Johnson at Research, Development, and Engineering Command (RDECOM) for sharing with us his work in autonomous flight operations. We also received valuable assistance from Colonel Robert Sova and his staff at the U.S. Army UAS Center of Excellence, from Mr. Ellis Golson and his staff at the Capabilities Integration Division, and from Colonel Joe Jellison and his deputy Mr. Glenn Harrison at the Directorate of Concepts and Requirements at the U.S. Army Aviation Center of Excellence. Thanks, too, to Allen Huber and Jim Towe at the Aviation Maneuver Battle Lab, to Michael J. Hahn, the UAS Futures Branch Chief, and to Tim Healy, the Operations Officer at U.S. Army Training and Doctrine Command (TRADOC) System Manager (TSM-UAS). At CASCOM, we were privileged to have the wise counsel and guidance of Ms. Andrea Jansen, Ms. Christine Myers, and CW4 Quitman Jackson. At RAND, our thanks go to the former program director, Eric Peltz, for his support for the project and guidance. We are also grateful to our associate program director, Rick Eden, for his help with the final report. Two of RAND's Army Fellows, LTC William Phillips and LTC Eloy Cuervas, were especially helpful in facilitating interviews with appropriate soldiers and units, and in providing their own insights into the project's issues. Finally, we thank our colleagues Randall Steeb and Elliot Axelband for their thoughtful reviews of an earlier version of this monograph.

Acronyms

AIB	Accident Investigation Board
ARSS	Autonomous Rotorcraft Sniper System
ASR	Alternate Supply Routes
BCT	Brigade Combat Team
BDOC	Base Defense Operations Center
BSB	Brigade Support Battalion
C4I	Command, Control, Communications, Computers, and Intelligence
C4ISR	Command, Control, Communications, Computers, Intelligence, Surveillance, and Reconnaissance
CAP	Combat Air Patrol
CAS	Close Air Support
CASCOM	Combined Arms Support Command
CBRNE	Chemical, Biological, Radiological, Nuclear and Explosive
CIDNE	Combined Information Data Network Exchange
CLS	Contractor Logistics Support
COCOM	Combatant Command

C-RAM	Counter-Rocket, Artillery, and Mortar
DARPA	Defense Advanced Research Projects Agency
DIA	Defense Intelligence Agency
DoD	Department of Defense
DOTMLPF	Doctrine, Organization, Training, Materiel, Leadership And Education, Personnel and Facilities
DZ	Drop Zone
EFP	Explosively Formed Penetrator
EO	Electro-Optical
EOD	Explosive Ordnance Disposal
FAA	Federal Aviation Administration
FLIR	Forward Looking Infrared
FM	Field Manual
FMV	Full Motion Video
FOB	Forward Operating Base
FSB	Forward Support Battalion
FY	Fiscal Year
GIS	Geographic Information Systems
GLO	Ground Liaison Officer
GMTI	Ground Moving Target Indicating (Radar)
HART	Heterogeneous Airborne Reconnaissance Team
HMMWV	High Mobility Multipurpose Wheeled Vehicle
HUMINT	Human Intelligence
IED	Improvised Explosive Device
IR	Infrared
ISR	Intelligence, Surveillance, Reconnaissance

JIEDDO	Joint Improvised Explosive Device Defeat Organization
JWICS	Joint Worldwide Intelligence Communications System
KTAS	Knots True Air Speed
LEMV	Long-Endurance Multi-INT Vehicle
LIDAR	Light Detection and Ranging
LOC	Line(s) of Communication
LSA	Logistics Support Area
MAJIIC	Multi-Sensor Aerospace-ground Joint Intelligence, Surveillance, Reconnaissance Interoperability Coalition
MP	Military Police
MPH	Miles per Hour
MSR	Main Supply Route
MTS	Materiel Tracking System
NASA	National Aeronautics and Space Administration
NBC	Nuclear, Biological, Chemical
NGIC	National Ground Intelligence Center
NIIRS	National Imagery Interpretability Rating Scale
OPSEC	Operations Security
O&S	Operating and Support
OSD	Office of the Secretary of Defense
QRF	Quick Reaction Force
RDECOM	Research, Development, and Engineering Command
RFID	Radio Frequency Identification

RPG	Rocket-Propelled Grenade
RSTA	Reconnaissance, Surveillance, Target Acquisition
SAR	Synthetic Aperture Radar
SIGINT	Signals Intelligence
SIPRNET	Secret Internet Protocol Router Network
TASS	Tactical Automated Security System
TF	Task Force
TTP	Tactics, Techniques, and Procedures
UAS	Unmanned Aircraft Systems
UAV	Unmanned Aerial Vehicle
UGV	Unmanned Ground Vehicle
USD(I)	Under Secretary of Defense for Intelligence
USTRANSCOM	U.S. Transportation Command
VBIED	Vehicle Borne Improvised Explosive Device
WISTI	Wide-Area Infrared Surveillance Thermal Imagery

CHAPTER ONE

Unmanned Aircraft Systems for Logistics Applications

Expectations for unmanned aircraft systems (UAS) run high within the Department of Defense (DoD) and the Army. Technological progress in propulsion, autonomous operations, sensors, weapons, and miniaturization of many components, buoyed by recent operational successes for UAS in Iraq and Afghanistan, has prompted significant interest in developing new applications for UAS that reduce manpower demands, reduce the risk to humans in the battlespace, or produce other benefits. The Army logistics community is no exception. It, too, has begun to think seriously about where UAS might be introduced into its operations to good effect. This monograph provides the results of a RAND Arroyo Center effort to identify and evaluate potential logistics applications (with the exception of resupply, which is being studied separately) for UAS out to the year 2024.

Relevant Studies

The research team's understanding of the feasibility of potential applications for UAS benefited from earlier research. A report released through the U.S. Army Training and Doctrine Command describes the Army's vision for the employment of UAS and makes clear how central these systems will become to a multitude of future Army operations.[1] Chow et al. (2009) identified ways UAS might work with fixed cameras and other sensors to secure logistics depots and infrastructure. Peters, Bonds, and Fischbach (2010) and Bonds et al. (2010) offered insights about the potential of future Army networks, including UAS, to protect logistics convoys, to detect improvised explosive devices (IEDs), and to foil ambushes through timely surveillance and reporting. These two reports emphasized the need for connectivity and appropriate receivers (e.g., terminals, computers, laptops, etc.) to enable all consumers of battlefield information to have low-latency, accurate situational awareness. In unpublished research for the Army on estimating the life-cycle cost of the multipurpose MQ-1C system, Peters et al. (2009) emphasized the importance of integrating UAS-borne sensors with other intelligence-collection disciplines and fusion in order to generate maximum results.

UAS Concepts for Evaluation

Concepts for evaluation were identified by the research sponsors, Army research and development personnel, and industry representatives. The research sponsors nominated a number of concepts for employing UAS for convoy and route security.[2] Others requested examination of the potential value of UAS to monitor pipelines and electrical lines to prevent sabotage. Staff at Research, Development, and Engineering Command (RDECOM) described their efforts to perfect autonomous rotorcraft that might prove valuable in retrograding critical unserviceable items. Industry suggested that several UAS might enable casualty evacuation under especially demanding battlefield conditions, such as those Israel experienced in its 2006 operations against Hamas and Hezbollah. Others identified potential capabilities such as predeployment

---

[1]   U.S. Training and Doctrine Command, U.S. Army UAS Center of Excellence, *"Eyes of the Army": U.S. Army Roadmap for Unmanned Aircraft Systems, 2010–2035*, no date.

[2]   Headquarters, USA CASCOM, *Logistics Re-Supply Mission Support Role for the Unmanned Aircraft System (UAS) Requirements Identification and Definition*, Information Paper, January 31, 2008.

reconnaissance to help logisticians plan the theater logistics infrastructure laydown and main supply routes in a new theater of operations.

Figure 1.1 summarizes the candidate logistics applications for UAS. The leftmost category, resupply, was the subject of a separate study undertaken by General Dynamics, so at sponsor direction, the RAND Arroyo Center effort focused its attention on the remaining three major categories: reconnaissance and surveillance (which might alternatively be thought of as protecting logistics convoys and assets), finding supplies, and transporting things.

Reconnaissance and surveillance has eight task subcategories. For the first, convoys, the research team considered whether convoy overwatch might be an appropriate response to elevated threat conditions.[3] For the second, route reconnaissance and surveillance, we considered whether UAS coverage of the route (as opposed to orienting on convoys moving along it) might improve the security of logistics convoys driving dangerous routes. For the third and fourth subcategories, surveillance of electrical lines and pipelines, the research team explored the benefits of having UAS monitor these linear targets to detect attacks and otherwise monitor the well-being of these important utilities in a combat zone. In the fifth, river surveillance, we examined the potential to determine a river's navigability by deploying UAS with appropriate sensors. In the sixth, surveillance of supply depots, the research examined the potential role of UAS in a security role for these installations. For the seventh, surveillance of domestic disaster scenes, we considered the potential benefits of having UAS overhead in the aftermath of a disaster to help determine the extent of the damage and the types of contamination that might be present. The eighth, theater reconnaissance, involves how UAS might help logisticians plan the "theater laydown"—that is, the distribution of logistics assets and capabilities—before the U.S. military enters or embarks upon operations there.

---

[3] Overwatch implies two units working cooperatively. The first takes up a supporting position from which it can observe the second unit's movements, warn of impending danger, and provide fire support. Overwatch, therefore, implies armed surveillance within supporting distance of the moving unit. See Headquarters, Department of the Army, *Mechanized Infantry Platoon and Squad (Bradley)*, Washington, D.C., Field Manual 3-21.71, August 20, 2002, Chapter 3 and Appendix A for full details.

Figure 1.1
Candidate Logistics Applications for UAS

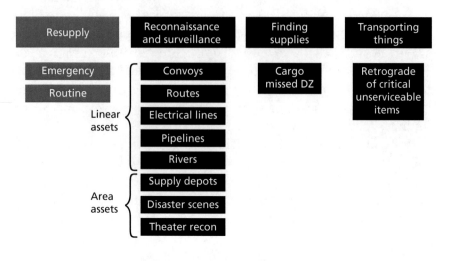

Within the next category, finding things, we initially examined the value of UAS for logistics inventory control purposes and for locating airdropped cargo that missed the drop zone. The research in the final category, transporting things, examined the potential of UAS to perform retrograde of critical unserviceable items.

Research Methodology

As originally conceived in the project description, each prospective UAS role would be evaluated through a three-step process: (1) assess its technological feasibility and development timing, (2) assess its likely cost, and (3) compare it to other non-UAS solutions. However, other factors emerged that led us to modify the research approach. Operational feasibility became a major consideration because each UAS application would take place in a military theater or contingency where the specifics about the enemy, weather, terrain, the strength and disposition of friendly forces, and other factors could be key determinants of a UAS concept's feasibility. For example, a UAS might be found quite capable of monitoring a pipeline: its sensors provide the resolution necessary to determine whether men seen near the pipeline are armed or not, the UAS is fast enough to revisit points along the pipeline's length at acceptable time intervals, and the UAS has the endurance to stay aloft over the pipeline for some acceptable number of hours. But if other factors that are key to the success of the surveillance concept are not favorable, such as the enemy's time near the pipeline is only fleeting, there are no friendly units close by to render timely intervention, or rules of engagement do not allow indirect fires, the concept will not work.

The research design that the team ultimately employed had four major elements: (1) technical feasibility, (2) operational feasibility, (3) benefit analysis, and (4) cost analysis. Figure 1.2 illustrates the process as a decision tree.

Figure 1.2
Research Design Decision Tree

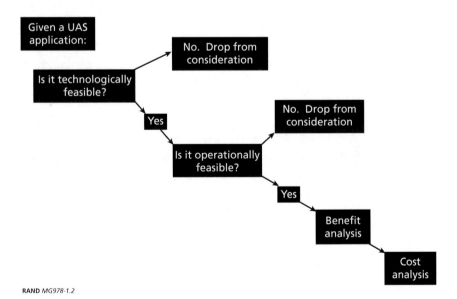

Technical Feasibility

We asked several basic questions to determine technical feasibility.

- First, do industry forecasts indicate that the necessary technology will be ready within the timeframe considered in the study, 2015–2024?
- Second, is there evidence of analogous technology in use today, perhaps in civil industry, academia, scientific circles, or elsewhere, that might suggest the technology would become feasible for logistics applications in the near term?
- Third, are there foreign examples of applications that might offer similar indications of future feasibility for logistics purposes?
- Finally, are key technologies at appropriate technology readiness levels (TRL)? In the course of the analysis it became clear that TRL was not a factor, so no further mention of this appears.

Operational Feasibility

Determining operational feasibility rested on five criteria: enemy, weather, terrain, number and disposition of friendly forces, and the size, frequency, and scale of logistics operations contemplated. Enemy considerations included the threat posed to logistics operations, consisting of the types of attacks, frequency of attacks, and severity of attacks. Weather factors included winds that might prohibit flight for certain classes of UAS, weather conditions that might interfere with sensor operations or data links, and icing that would interfere with flight for certain classes of UAS. Terrain considerations included mountains that might break electronic line-of-sight and require satellite uplinks to maintain data links and control of UAS, altitudes that might be beyond the performance characteristics for certain classes of UAS, and density-altitude problems (i.e., high altitudes and hot temperatures that reduce aircraft performance) that might preclude UAS operations at some altitudes under some weather conditions.

The number and disposition of friendly forces considerations included the force-to-space and force-to-population ratios in a given theater, the relative permissiveness of the environment, typical response times for quick reaction forces (QRF), the availability and responsiveness of indirect fires, the rules of engagement, and the general responsiveness that logisticians and logistics units might expect from the combat formations in coming to the assistance of logistics formations.[4]

Benefit Analysis

In the benefit analysis, we developed two distinct dimensions. The first was to identify the specific benefits associated with the UAS application under investigation. Does the UAS application:

- Reduce manpower requirements?
- Remove personnel from dangerous jobs or circumstances?
- Improve force protection?
- Reduce materiel requirements associated with the operation or task at hand?
- Accelerate the tempo of operations in useful ways?
- Reduce costs?
- Simplify processes?
- Reduce the likelihood of attack?
- Reduce the costs incurred from a successful attack?

The second dimension of benefit analysis had to do with comparative benefits. Here the questions turned to whether the UAS-based approach was superior to non-UAS-based alternatives, including today's practices. For example, when it comes to UAS providing surveillance of a logistics support area, is the UAS performance superior to that of fixed cameras deployed for the same task? Which approach provides superior coverage of the installation? Which approach provides the highest-resolution images? Which provides the longest-range intrusion detection?

---

[4] Force-to-space ratios have to do with the relative abundance of U.S. combat forces given a country of a certain size. Force-to-population ratios capture a similar relationship between combat forces and the indigenous population. Adequacy of U.S. forces understood through these ratios is often seen as critical for success in stability operations and counterinsurgency. See James Quinlivan, "Force Requirements in Stability Operations," *Parameters*, Winter 1995, pp. 59–69. These ratios interact with "permissiveness."

Cost Analysis

The cost analysis contained absolute and relative cost dimensions: What would the UAS application itself cost? How do those costs compare with the costs for non-UAS-based alternatives and compared to today's practices?

Presentation of the UAS Logistics Concepts

The remainder of this report is organized around the broad categories of UAS tasks illustrated in Figure 1.1. Chapter Two places the analysis in the context of rapidly evolving technological and operational changes that may affect conclusions based on current experience and capabilities. Chapter Three summarizes each of the concepts and the analytical conclusions about their technical feasibility, operational feasibility, cost, and benefits (a detailed examination is in Appendix A). Each UAS logistics concept begins with a description of the fundamental job itself (e.g., providing convoy security). The description explains how the job is performed in operational theaters today (Case Zero), how it might be performed with UAS (Case One), and how it might be performed with an alternative, non-UAS-based solution (Case Two). Chapter Four offers observations about concept implementation, suggesting how the logistics community might establish priorities among the most attractive applications, and how it might go about acquiring UAS for selected applications by competing for use of UAS currently in the force structure, as well as other approaches. Chapter Five provides the project's observations and conclusions. Appendix A presents a more detailed analysis of the individual concepts for the use of UAS. Appendix B describes the issues surrounding sensors and imaging from UAS. Appendix C details the analysis of cost and benefits associated with the UAS concepts examined. Appendix D provides an overview of current UAS. Appendix E offers recommendations for handling infeasible and underperforming concepts.

CHAPTER TWO

The Near-Term and Long-Term Future

Expectations for the Future

This chapter considers the future, both the next few years and out to 2024. Doing so is fraught with uncertainty, because Army plans for UAS and C4ISR may change rapidly as the service learns from its ongoing operations and as new developments demand new solutions. Issues that only 12 months ago seemed intractable—Federal Aviation Administration (FAA) constraints on UAS operations in U.S. airspace, for example—are being resolved. Predictions under these circumstances are likely to be wrong or badly off the mark. But we can consider some of the trends that are in evidence today and explore what they suggest about the near- and longer-term future, knowing all the while that variables not yet in evidence may intervene and change things.

Sensitivity to Cost and Benefit Developments

The cost-effectiveness of the UAS concepts examined in this monograph is sensitive to three variables: the cost of the UAS and their payloads, the effectiveness and size of sensors, and the cost of the damage sustained by logistics formations that the UAS would have precluded. Analysis reveals positive trends for all three. The advent of less expensive class II UAS (e.g., Scan Eagle) and the trend in miniaturization of sensors offer the potential of high-benefit, low-cost UAS in the near term. As sensing technologies undergo successful miniaturization, the prospect of high-resolution sensing becomes more likely, and with higher resolution, reconnaissance and surveillance applications for UAS are likely to benefit. As for the cost of damage, if a future adversary can sustain a high frequency of attacks on logistics convoys and if each such attack results in significant materiel losses and friendly personnel killed in action, these costs can produce a business case for a given UAS application.

For the reconnaissance and surveillance concepts (armed convoy overwatch in which an armed UAS shadows a convoy; route surveillance (Army) where UAS are employed to maintain a view of a line of communication (LOC); route surveillance (Joint Fires), which employs a combination of surveillance systems and response options; and finally, fixed surveillance, which deploys tower-based sensors), Figures 2.1 through 2.3 illustrate how the benefit of employing UAS depends on the cost of the systems themselves versus the cost of the damage they can prevent. Figure 2.1 shows the base case.

Figure 2.1 is derived from significant activities reports from Iraq during the July 2009 timeframe. The x-axis reflects the probability of

Figure 2.1
Base Case Benefit Analysis

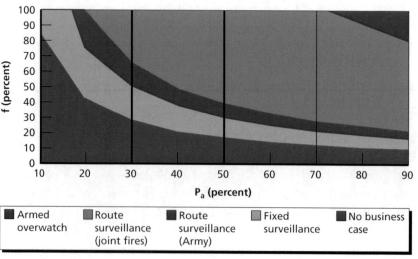

attack, and the y-axis reflects as increasing effectiveness (by avoiding IEDs, ambushes, and similar attacks), represented as a percentage, of Army detection capabilities. We estimated that logistics formations incur between 30 and 50 percent of the attacks. The brown area in the figure reflects the region in which the UAS examined in the study do not produce sufficient benefit to warrant acquiring them. For example, if the probability of attack were 20 percent and UAS increased detection effectiveness by 30 percent, the UAS would not be cost-effective. The tan-colored band indicates the amount of additional effectiveness for a given probability of attack necessary to warrant fixed surveillance solutions. The green band reflects the additional effectiveness necessary to warrant UAS for route surveillance. Note that if logistics units were incurring attacks with a high probability, then less effectiveness is needed to warrant the use of UAS for each of these roles.

Figure 2.2 shows the effect of cheaper UAS. If UAS were half the cost of the ones examined in this study, the benefit analysis changes significantly. Such a cost reduction may be possible if the Army moves

Figure 2.2
UAS Half the Cost of Systems Considered

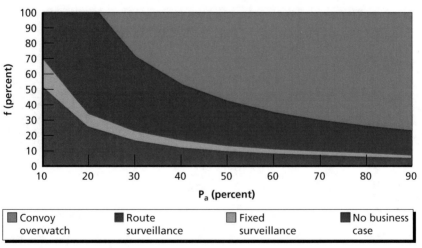

toward class II UAS. The appearance of "mini-SAR"—greatly miniaturized synthetic aperture radar—is indicative of the ongoing trend in
new, cheaper, smaller, high-quality sensors.

Similar changes occur if damage increases significantly. Figure
2.3 illustrates the effect on the benefit analysis if twice the amount of
damage is considered.

This analysis makes the point that the value or cost-effectiveness
of UAS are sensitive both to system cost and to the cost of damage the
system can prevent. Future circumstances and the availability of less-
expensive UAS, especially class II systems, could make UAS attractive
for the applications in question, in terms of both feasibility and benefit.

The Near Term

A number of developments suggest near-term technology improvement potential. Several indicate potentially important improvements
in reconnaissance and surveillance, including DARPA's Heterogeneous
Airborne Reconnaissance Team (HART) and the Air Force's "Gorgon

Figure 2.3
Twice the Damage

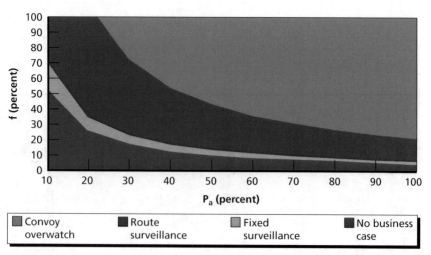

Stare." HART will provide persistent operations on a continuous basis, exploiting multiple sensors aboard multiple aircraft—both manned and unmanned—to provide "video-on-demand from multiple sources to multiple users."[1] The system is expected to have capability for geo-registered multi-sensor mosaic images. Gorgon Stare offers the prospect of a 12-camera array aboard Predator and Reaper UAS capable of concurrent imaging over a four-kilometer radius. The multiple look-angles associated with the cameras are expected to be revealing of details that single-camera sensors do not capture.[2] DARPA's "ARGUS" program may attempt a more ambitious, similar effort with 92 individual cameras.[3] If these efforts develop along the lines anticipated by their program managers, users should receive higher-resolution imagery than that available with today's sensors.

Industry is responding vigorously to the DoD's need for small, more capable UAS, producing many potentially useful candidates. For example, PUMA, a slightly larger, hand-launched, battery-powered Raven-like UAS, has debuted with a better sensor suite, longer mission endurance, and wider surveillance range than the original Raven. Its minimal visual profile and quiet engine allow it to fly at low altitudes that optimize its sensors' performance while remaining difficult to detect by those on the ground beneath its orbit.[4]

Some sensing technologies are also yielding improved performance. Synthetic aperture radar systems have evolved to produce imaging more quickly (2–4 minutes) than earlier generations and can, under some circumstances, produce images of four-inch resolution. Some, including the Lynx SAR, can be programmed to follow a road or similar terrain feature. These sensors feature scalable levels of resolution and a "spotlight mode" that allows the sensor to linger on the item

---

[1] Dr. Michael A. Pagels, DARPA, "Heterogeneous Airborne Reconnaissance Team (HART)," briefing, August 2008.

[2] Stephen Trimble, "USAF to Unleash 'Gorgon Stare' Sensor in 2010," *Flight International,* January 28, 2009.

[3] Noah Shachtman, "Air Force to Unleash 'Gorgon Stare' on Squirting Targets."

[4] For details, see http://www.designation-systems.net/dusrm/app4/puma.html

of interest. They also operate in ground moving target-indicating mode (GMTI) and are sensitive enough to detect small vehicles.[5]

Not all of the factors influencing the near-term outlook for UAS are positive, however. Bandwidth remains a concern. A Congressional Budget Office study raised the issue in 2003, noting the growth of demand for bandwidth at each echelon of command from platoon through corps as the Army continued to develop its C4ISR architecture.[6] A subsequent RAND Arroyo Center report studied the issue and concluded that demand for bandwidth will exceed the supply, that bandwidth must be managed and allocated as an important combat resource, and that no single approach would solve the Army's bandwidth problem.[7] The proliferation of intelligence sensors aboard all platforms and the resulting data they collect and disseminate create new demands for bandwidth.

Enemy countermeasures are an emerging concern. It has been reported that militants in Iraq have been able to use cheap, readily available technology to intercept live video feeds from Predator UAS, that video from Predators has been found on captured enemy computers, and that Iraqi insurgents seem to be learning how to identify an area under surveillance and avoid it.[8] Data encryption may solve the immediate problem, but if this practice evolves into a broader cyber-war capability for countering U.S. UAS, the Army and the DoD might find the value of certain classes of UAS degraded, which might, under worst-case circumstances, undermine the U.S. tactics, techniques, and procedures (TTPs) that the UAS support.

Conditions like those in Afghanistan could also degrade UAS effectiveness. On the positive side for UAS, Afghanistan has far fewer

---

[5]   For a complete description of Lynx performance, see Jane's Electronic Mission Aircraft, "Lynx," June 18, 2010.

[6]   Congress of the United States, Congressional Budget Office, "The Army's Bandwidth Bottleneck," August 2003.

[7]   Leland Joe and Isaac R. Porche, III, *Future Army Bandwidth Needs and Capabilities*, Santa Monica, Calif.: RAND Corporation, MG-156-A, 2004.

[8]   Siobhan Gorman, Yochi J. Dreazen, and August Cole, "Insurgents Hack U.S. Drones," *Wall Street Journal*, December 17, 2009, p. 1.

paved roads than Iraq. Gravel roads and graded trails make for slower convoy speeds, which may give convoy commanders more time to react to UAS-provided data. On the other hand, poor road conditions can also undermine sensor performance, because gravel and unpaved tracks may produce clutter that interferes with the sensor as road traffic lofts dust and debris into the air. Weather is another consideration. In some parts of Afghanistan, for example, weather conditions are so difficult that Hunter UAS are flyable less than four months per year.[9] Rain and snow can also degrade the performance of some sensors.

Despite these concerns, the near-term evolution of UAS and their sensors are likely to produce options that cost less and offer better imaging than current UAS, which were used for the base case analyses. If such opportunities emerge, the Army may want to employ them in a way consistent with the logic illustrated in Figures 2.1 through 2.3—that is, as one course of action among several, to be employed when the probability of attack and the potential cost of an attack appear to warrant it.

2024: The Longer-Term Future

The connection between current developments and the longer-term future may not be as firm as the link between recent advances and the near term; certainly the longer timeframe leaves more opportunities for unanticipated circumstances to deflect the longer-term future from the trajectory we might otherwise project or posit. That said, Lieutenant General John C. Koziol, the Deputy Under Secretary of Defense (Intelligence), has a vision for the future that may subsume individual UAS programs, manned programs, and space-based systems into a robust network where computers rather than humans are the network nodes, and where information of all sorts is available on demand.[10]

His office has begun buying some of the key parts of the network. These include the Battlefield Airborne Network Node (BACN), which will serve both as a communications relay and a network gateway. The

---

[9] Email exchange between RAND analyst Kenneth Horn and Major Jon Beale of the Army's 801st Brigade Support Battalion (BSB), December 14, 2009.

[10] Based upon General Koziol's presentation at RAND, November 5, 2009.

Long-Endurance Multi-INT Vehicle (LEMV) is an unmanned, long-endurance hybrid airship capable of deploying a wide array of various sensors.[11] This airship is expected to operate soon in Afghanistan.

Assuming that General Koziol's vision comes to pass, the resulting network would feature constellations of information-producing systems including satellites, UAS, and manned platforms operating at various altitudes, carrying different sensor payloads. According to this vision, the network would eventually be capable of machine-to-machine collaboration and multi-INT fusion. It would produce multi-INT mosaics and similar products, potentially becoming a theater-level version of DARPA's HART with a richer suite of sensors and platforms.

Using the network might then be akin to watching satellite television. For example, a commander and his staff planning an operation in Paktika province, Afghanistan, might tune in to channel 512 for current imaging of the terrain their unit will traverse, employing change detection to identify potential IED sites to be avoided. A convoy commander about to move up main supply route (MSR) Tampa in Iraq might turn to channel 830 for the latest data on his route. The duty team in the base defense operations center at a logistics support area (LSA) or forward operating base (FOB) could likewise watch channel 1230 for video of their perimeter and the surrounding area. An attack helicopter responding to a perimeter attack might look at channel 1230 to find the intruders, then switch to its own forward-looking infrared (FLIR) to fight the engagement. As with satellite TV, anyone with a subscription could watch any channel; thus, Army personnel doing very different jobs might nevertheless be employing the same channels on the network to meet their own reconnaissance and surveillance requirements.

General Koziol's vision has implications for all military entities currently considering or actively seeking UAS. First, distinctions between "strategic" and "tactical" assets lose meaning as progressively larger Army audiences can exploit information derived from some combination of these systems for their own mission needs. Second,

---

[11] See Graham Warwick, "Staying Up, Staring Down, LEMV Airship," in "Ares—A Defense Technology Blog," *Aviation Week*, June 8, 2009.

distinctions between "general support" systems and "direct support" systems will become progressively less important as satellites, hybrid airships, UAS, and manned systems become more abundant because, as their numbers in theater grow and the data they collect become integrated with other sources of information (perhaps many times in a very short period of time), the probability of useful imaging appearing on a given channel increases. Just as fans can program Monday Night Football with their satellite TVs before they leave for work in the morning and that evening catch the opening minutes of the first quarter of play they missed while stuck in traffic, future Army units may be able to program their intelligence needs in advance and rely on the network to download the appropriate data for them.

As the sensors on the network become more abundant and theater coverage becomes both denser and higher in quality than what we experience today, unit ownership of specific sensors becomes less critical—until the point at which the network can no longer provide bandwidth to support all of the would-be subscribers. At that point, the Army will want to reconsider its options. One option might be to buy simple, cheap, small UAS with portable viewer/control stations and distribute them to priority units. These UAS would make few demands on the network and its bandwidth because their data would stream to a single point, the operator on the ground. This course of action is not the obvious or only one, however. Today the Army lacks sufficient bandwidth in Afghanistan, faces conditions that often do not favor small UAS (weather, clutter), but manages to maintain its freedom of movement and tactical supply stock levels at acceptable risk by employing a mix of "jingle air" (Mi-8 helicopters and small, fixed-wing aircraft flown by contractor air crews), "jingle trucks" (locally contracted trucks), and "green air" (U.S. Army aviation, typically CH-47s, though not exclusively) to move materiel and manpower around the area of operations.

The big question is bandwidth and whether the Army can create enough of it to support the network that lies at the center of General Koziol's vision. If bandwidth is estimated to be sufficient, then logisticians preparing for 2024 should buy terminals so that logistics units of that era can connect to the network and all of the sensors—ground, air and space, manned and unmanned—in it. If bandwidth is estimated to be insufficient, then logisticians, like their counterparts in other branches, should begin developing their priorities in anticipation of rationed bandwidth. Investing in Army-owned UAS would provide a hedge against the risk of insufficient bandwidth to support the envisaged network of sensors.

CHAPTER THREE

Examining the UAS Concepts

This chapter provides summary analyses of the potential UAS applications that are the centerpiece of the project's research. A more detailed treatment appears in Appendix A. Each section briefly describes the UAS concept, its technical and operational feasibility, and an overview of its cost and benefits. Detailed cost and benefit analysis appears separately in Appendix C.

Convoy Overwatch

Convoy overwatch is an adaptation of a maneuver technique for logistics purposes. Traditionally, overwatch involves one formation advancing while another observes and monitors its progress.[1] When the overwatching (monitoring) unit sees trouble in the path of the advancing unit, the overwatching unit provides suppressive fire and intelligence to help the advancing unit continue its mission either by avoiding the awaiting enemy ambush or by destroying the enemy. The notion of convoy overwatch envisions an armed UAS flying in direct support of a logistics convoy, using its real-time sensors to help the convoy commander detect threats along the road ahead, and having the UAS eliminate them before the convoy arrives.

---

[1] See Headquarters, Department of the Army (2002), Chapter 3, for full details.

Technical Feasibility

It is technically feasible to fly an armed UAS with real-time sensors in support of a convoy. Armed UAS are commonplace in today's operations. The potential benefit of convoy overwatch depends on the capabilities of the available real-time sensors (sensors whose output can be understood without time-consuming post-processing and annotation, making the output available almost immediately as the sensor senses the phenomena it is designed for). Today's real-time sensors can detect about one-third of the threats to logistics convoys. They do not generally "see" IEDs, detect snipers, and similar threats (e.g., grenades thrown from crowds) that endanger today's convoys with sufficient fidelity to identify them.[2]

Operational Feasibility

Operational feasibility also reveals limitations in using UAS for convoy overwatch when it operates at relatively high speeds. Logistics convoys in Iraq travel at around 40 miles per hour (MPH).[3] At such speeds, a UAS attempting overwatch would have to be well out in front of the convoy (creating a gap that the enemy could use to lay an ambush) in order to afford the convoy commander sufficient time to identify a threat from the UAS data, form a course of action, transmit it to the march elements in the convoy, and execute it.[4] Moreover, if the convoy commander needs fire support from the UAS, then he must negotiate the six-step process for targeting, which could consume more time.[5]

Because of the number of logistics convoys, using UAS pervasively for overwatch in Iraq at the height of the insurgency would have

---

[2] See Appendix B for an analysis of sensors and their limitations.

[3] Interviews with 311th Expeditionary Sustainment Command personnel, March 30 and June 5, 2009. Speeds are slower in Afghanistan, especially on unimproved roads and trails. See Appendix A for details.

[4] The UAS would have to be 536 meters ahead if the convoy commander can accomplish these steps in 30 seconds, 1,073 meters ahead if he needs 60 seconds, and 1,609 meters ahead if he needs 90 seconds.

[5] The steps are find, fix, track, target, engage, and assess. See Joint Chiefs of Staff, *Joint Targeting*, Joint Pub 3-60, April 13, 2007.

required a substantial increase in UAS investment. Consider that typically 19–23 convoys left Kuwait each day to resupply the FOBs in Iraq. Convoys traveling up MSR Tampa may take seven days to complete the trip. Tasking UAS to provide overwatch for each convoy for up to seven days would require use of most of the armed UAS sorties available in theater today. Appendix C provides detailed estimates of the costs of sufficient UAS to conduct overwatch in Iraq.

Circumstances may evolve differently in Afghanistan or future theaters, but the fundamental considerations are the same: the number of convoys that must run, the number of UAS sorties needed to cover them, the value of the damage and casualties avoided per UAS patrol, and the cost of the UAS themselves relative to the available cheaper, non-UAS-based alternative solutions for convoy protection of similar efficacy. If the end-to-end costs of the UAS necessary to accomplish the mission are less than or equal to the costs of the damage and casualties avoided, and cheaper alternatives of similar efficacy are unavailable, then UAS would be the tools for the task. Technology may make these conditions easier to satisfy as current sensing phenomenologies become capable of higher resolution and new sensing technologies appear on the scene. Miniaturization of sensors and the acquisition of smaller, cheaper UAS such as the class II aircraft could reduce costs. Conditions like those in Afghanistan tend to remove alternative solutions from the equation. For example, the sheer length of the national ring road renders ground camera–based surveillance infeasible, and line-of-sight problems associated with the rugged terrain tend to favor aerial surveillance.

Route Surveillance[6]

Route surveillance orients the UAS on the road rather than on the convoys moving along it. The analysis here examined adding capabilities

---

[6] A systems approach to route/MSR/LOC surveillance might include placing UAS in the appropriate density under the authority of the units tasked with route security. The units in question would operate UAS over the priority routes. They could provide the resulting full motion video or imagery over laptop to convoys (and other route users) as an imagery adjunct

to the way the job is typically performed in Iraq.[7] Case One considered three MQ-1C UAS conducting surveillance of MSR Tampa covering about 20 percent of the route at any time, with two-hour gaps between surveillance passes. Close air support was assumed to be available with a 30-minute response time. Case Two examined deploying fixed cameras and four light utility helicopters to augment the Military Police (MP) battalion with route security responsibilities. The MPs would man three outposts at 75-mile intervals to protect the mast-mounted cameras. They would have four light utility helicopters to supplement the cameras and a QRF available within 30 minutes.

Technical Feasibility

As with convoy overwatch, the benefit of UAS-based route surveillance in Case One depends on the capabilities of available real-time sensors. The more fidelity offered by the sensor, the greater the ability to detect the most dangerous threats to convoys. Currently the approach is technically feasible (UAS fly similar missions for maneuver units all the time), although the current sensors cannot achieve a National Imagery Interpretability Rating Scale (NIIRS) value sufficient to reveal IEDs or other forms of hostile intent reliably. Future sensors may overcome the current limitations either by deploying higher-resolution systems than are possible today or by employing new sensing phenomenologies that reveal greater detail.

Case Two, deploying fixed cameras from three strong points, has the advantage that fixed cameras have already achieved: high-fidelity imaging. Fixed cameras have been demonstrated to produce high-fidelity images over relatively long distances, subject to degradation

---

to the voice communications over Sheriffs Net to improve the amount, quality, and timeliness of information about threat conditions along LOCs in the theater.

Alternatively, the sustainment brigade responsible for the resupply convoys might be given responsibility for operation of the entire system. Such an arrangement would produce unity of effort and parsimony while making the imagery available to all route users.

[7] Today, typical route security missions fall to MP battalions, cavalry squadrons, and engineer battalions. Brigade combat teams (BCTs) may task their maneuver formations with route security under some circumstances.

by weather, darkness, and enemy deception. Fixed cameras typically outperform their airborne counterparts.[8]

Operational Feasibility

The operational feasibility of this potential application can be limited by enemy behavior to avoid detection, as illustrated by today's adversaries and their tactics. Most of the threats to logistics convoys are fleeting in nature and do not present a clearly hostile posture until the final instant of the attack: the explosion, the bullet's impact, etc. IED teams do not typically linger at the roadside to plant their device. Indeed, they may make multiple trips to install different components, limiting their vulnerability to detection during any one step of the process. Grenades are often thrown from crowds, dropped off overpasses, or thrown from passing cars and motorcycles. There is, therefore, little for sensors to see, and what there is to see lasts only for instants. The emerging science of "emblematics"—the ability to interpret involuntary physiological responses from a distance—may eventually provide new clues to reveal lurking hostile intent, but the subject is at present in its infancy, and it is difficult to anticipate progress to specific levels of capability.

Cost

Life-cycle cost estimates indicate that Case Two, fixed surveillance from strong points manned by MPs and supported by light utility helicopters, is the least expensive option for this mission at approximately $570 to $770 million in annual life-cycle costs. Case One, where UAS are used for route surveillance, is more expensive at approximately $0.8 to $1.2 billion for the same route, and would require a minimum of 16 UAS, or about half of the 31 combat air patrols conducted by armed UAS in Iraq during the course of this research.

---

[8]   See Brian G. Chow, John E. Peters, Katherine Comanor, Marvin Schaffer, and Edward R. Harshberger, *Fighting Air Bases Under Attack: Forward Operating Bases*, Santa Monica, Calif.: RAND Corporation, 2009 (not releasable to the general public), for details on various cameras.

Electrical Line Surveillance

Surveillance of electrical lines to prevent sabotage was one of the concepts nominated for the study. Upon investigation, however, it became clear that electrical system security is not a logistics function, but rather an engineering one. Thus, the sponsor directed that we not examine this application. Commercial infrastructure protection usually falls to the unit commanders in charge of a given area, and is typically realized through local security arrangements involving some mix of indigenous police and security and U.S. military forces. That said, electrical lines share many features and vulnerabilities with pipelines, discussed below.

Pipeline Surveillance

The concept examined envisions using UAS to provide surveillance of Army "temporary" pipelines such as those installed and operated by petroleum pipeline and terminal operating companies.[9] In the absence of sufficient MP units (i.e., Case Zero as it operates in Iraq), these pipeline companies often end up securing the pipelines themselves.[10] Because these pipelines are expeditionary, they are typically installed on the surface but have earth berms and a layer of concertina barbed wire for protection and security. The pumping stations that punctuate the pipeline (between 18 and 58 miles apart) are manned by approximately 12 soldiers each who look after the pipeline's security and its functionality.[11] Case One involves flying UAS at some frequency to provide surveillance. Case Two employs fixed cameras at the pumping stations to provide continuous coverage between them, along with

---

[9]   For protecting linear assets like pipelines, it may be useful to fly UAS at low altitudes where they can be seen and heard for their deterrent value. Many of our interviewees noted that the enemy avoids UAS when they know the aircraft are up. Small UAS like Ravens could be flown from pumping stations to add a layer of security.

[10]   David M. Oaks, Matthew Stafford, and Bradley Wilson, *The Value and Impacts of Alternative Fuel Distribution Concepts: Assessing the Army's Future Needs for Temporary Fuel Pipelines*, Santa Monica, Calif.: RAND Corporation, TR-652-A, 2009.

[11]   240th Quartermaster Battalion After Action briefing.

additional strong points between pumping stations (e.g., those 58 miles apart) to maintain near intervisibility between fixed cameras.

Technical Feasibility

Judging from current industry examples, flying UAS for pipeline surveillance should be technically feasible. We found examples in the oil industry, specifically with protection of offshore oil rigs, where UAS have been flown with FLIRs to secure the rigs against pirates and terrorists. The sensors seem to be effective, with efficacy varying with environmental conditions.

Operational Feasibility

Operational feasibility is contingent upon the threat level (how intensive are efforts to damage the pipelines?), the TTPs employed by the enemy (do they create a recognizable hostile signature that lasts long enough to be detected?), the length of the pipeline and the number and speed of the UAS devoted to protecting it (are there gaps in coverage or long revisit rates?), and the availability of appropriate responses (QRF, indirect fire, etc.).

These factors vary by situation. In Iraq, circumstances have limited the benefit of using UAS for pipeline surveillance: enemy attacks are relatively infrequent, present a small signature, and are of a short duration.[12] Moreover, the costs associated with pipeline attacks are relatively low, typically less than 20 percent of throughput. Finally, it would be difficult to act on UAS information, given the dearth of QRFs within responsive range of the pipelines and the impracticability of protecting pipelines with indirect fires.[13] In conditions where the costs and the consequences of pipeline damage were greater, then employing UAS for pipeline surveillance would be more operationally valuable.

---

[12] 240th Quartermaster Battalion After Action briefing. Interviews with representatives from Arctic Slope and Range Services (ASARS), who do defense contracting in Iraq for various services and functions.

[13] 200th Quartermaster Battalion After Action Report.

Cost

Cost is a function of the number of UAS employed, which in turn is a function of the pipeline length and required revisit rate.

River Navigability

For this concept, UAS would be employed to ascertain whether a river was navigable (i.e., free of mines and obstacles) and could, therefore, be exploited as a possible supply route. This is a highly specialized form of route surveillance, and its potential benefit depends on a specialized kind of sensor called a LIDAR (light detection and ranging).[14] Today's LIDARs are large and sensitive instruments that require monitoring by a technician to keep within operating parameters. They are, therefore, not yet suited for employment aboard an unmanned system. When LIDAR technology has advanced in stability and miniaturization, this concept may become technically feasible.

Surveillance of Supply Depots

The research team has extensive experience with supply depots and UAS, LSA Anaconda/Balad Airbase in particular. We will depart from our formalism of technical feasibility, operational feasibility, and cost in this case to draw on the research cited in Chapter One, noting that in the detailed analysis of that installation, radars and fixed cameras were found to provide primary situational awareness and that UAS had a beneficial role in this task when integrated into the overall installation defense and surveillance architecture.[15]

---

[14] The consensus within the remote sensing community at RAND and sustained by the findings from academic programs like the Columbia University LIDAR project is that LIDAR is the appropriate sensor for such an application. The sensor can be confounded by a muddy riverbed, the weather, or by silt hanging in suspension in the water.

[15] For fixed facilities, UAS may best be deployed with QRFs to track perpetrators of indirect fire attacks. Ravens or similar small UAS might be assigned to the installation QRF. When the Base Defense Operations Center (BDOC) detects incoming fire, the BDOC alerts the

UAS work best as part of a larger system, able to integrate their data feeds with those from other sources and sensors. Even in ISR applications, where UAS seem to have the most intuitive advantages, their performance is enhanced by integration with other intelligence-collection disciplines, specifically signals intelligence (SIGINT) and human intelligence (HUMINT), because the images themselves often do not reveal hostile intent in sufficient time for U.S. military personnel to take action. The imaging, informed by intelligence from other sources, often can establish that the men and vehicles in the images are indeed enemy combatants and satisfy the rules of engagement to take action against them.

UAS also have a useful role in installation security. Earlier research suggested they are more effective at pursuing perpetrators of indirect fire attacks and leading a QRF to them than other options. As sources of situational awareness for installation security, UAS are secondary to radars and fixed cameras. For other tasks, including the hunt for high-value individuals, there is no obvious alternative to the persistent and extensive use of UAS. But this is a resource-intensive application. For example, the 3d Special Operations Squadron reports that it takes ten hours of full motion video (FMV) or 13.4 UAS sorties for every high-value individual captured or killed.[16] Others require more resources. It took approximately 600 hours of UAS time to locate al Zarqawi, for example.[17]

Support to Domestic Disaster Responses

In this application, UAS would fly to assist first responders in surveying domestic disaster scenes and ascertaining their extent and nature.

QRF, which, as a first step, would launch a UAS in the direction of the incoming enemy fire. The QRF then would then monitor the data feed from the UAS to locate and follow the suspects until they are able to make an apprehension.

[16] Lt Col Bob Brock, USAF, "3d SOS Command Brief: UAS Vision and Frameworks," April 20, 2009. In these operations the UAS do not operate alone, suggesting that independent UAS missions would be even more time-consuming.

[17] Peters et al. (2009).

The FAA and Department of Defense have been directed to find a solution to the airspace access issue that has until recently rendered this concept technically infeasible.[18] The two departments currently have a memorandum of understanding that commits the FAA to process DoD requests for a Certificate of Authorization or Waiver (COA) to support UAS operations as part of the U.S. disaster response within 24 hours.[19]

Predeployment Theater Reconnaissance

The notion here is that logistics commanders would make use of high-altitude, long-endurance UAS to survey a future theater of operations; identify likely routes suitable as MSRs/alternate supply routes (ASRs); identify sites for future FOBs, fuel depots, and other logistics sites; and ultimately develop their plans to support the operations to come. Such missions would take place in addition to Case Zero planning, which ideally should make full use of contingency planning products: route studies, trafficability studies, operational support studies, joint theater transportation studies, and the U.S. Transportation Command (USTRANSCOM) Secret Internet Protocol Router Network (SIPRNET) site.[20] Logistics planners could also usefully consult the National Ground Intelligence Center (NGIC) Geographic Information Systems (GIS) data holdings for the region in question. Commer-

---

[18] Section 935 of the 2009 Defense Appropriations Act (H.R. 2647-247) requires that "the Secretary of Defense and the Secretary of Transportation shall, after consultation with the Secretary of Homeland Security, jointly develop a plan for providing expanded access to the national airspace for unmanned aircraft systems of the Department of Defense." According to the Association for Unmanned Vehicle Systems International (December 2009), a new, ground-based sense-and-avoid radar system is being developed to support prompter access to U.S. airspace for Army UAS.

[19] Memorandum of Agreement for Operation of Unmanned Aircraft Systems in the National Airspace System dated September 24, 2007 and signed by the Deputy Secretary of Defense and the Federal Aviation Administrator.

[20] https://intelink.intel.scott.af.smil.mil/components/JIC/index.cfm (via SIPRNET)

cially available satellite imagery from companies like Digital Globe, Google Earth, and Geosage might also prove valuable.

Case Two is the U.S. Geological Survey's WB-57 hyperspectral imaging support flown aboard NASA manned aircraft. This capability was authorized in National Security Presidential Directive 44 and endorsed in DoD Directive 3000.5, "Military Support for Stability, Security, Transition and Reconstruction Operations." The system can provide multi-spectral imaging for specific logistics issues and applications. It offers ease of access (because there is lower demand for this capability than for UAS), better resolution for logistics purposes, and multiple sensing options.

Technical Feasibility

Case One, the use of UAS in this role, is technically feasible. High-altitude, long-endurance platforms fly a variety of sensors that might provide useful information to logistics commanders. Specific utility to logistics planners will be situationally dependent.

Operational Feasibility

Operational feasibility is contingent on several considerations. First, the logistics community would need access to the data coming off the UAS or WB-57. This requirement suggests the need for SIPRNET or perhaps Joint Worldwide Intelligence Communications System (JWICS) terminals, depending upon the organizations flying the missions that produce the data.

Second, the logistics community would have to compete for priority access to the UAS that are flying. While logisticians and maneuver commanders may sometimes find the same information of interest, it is likely that the logisticians will seek to fill specific information gaps through UAS sorties. Thus, these logisticians will need sufficient priority over other claimants on the flight to ensure that the mission overflies their specific areas of interest with appropriate sensors on board.

The third consideration is operations security (OPSEC). Flying UAS into/over enemy airspace may be imprudent if it is likely to provide warning to the enemy that U.S. military operations may loom.

Finding Airdropped Cargo That Missed Its Drop Zone

Case Zero and Case One were combined in the examination of this application, and they involve deploying a search party to look for the cargo. Search parties would typically look short and long of the drop zone and off to the downwind side. The local commander might request air support in the form of either a helicopter or a UAS, especially if weather conditions suggest a large search area or if there is a chance that the cargo landed in wooded or other terrain that interferes with a search party's ability to locate it.

Case Two employs a 3G active radio frequency identification (RFID) tag on the cargo, which displays the cargo's location on the Army's Materiel Tracking System (MTS). The ground commander consults the MTS in his tactical operations center to get the coordinates of the cargo, and moves the search and recovery party to that location.

Technical Feasibility

The concept is technically feasible. The U.S. inventory includes many UAS that could fly the mission, and current sensors would be able to detect cargo parachutes and the palletized load. Detection might be enhanced by sewing reflective tape onto the parachute canopies to make them more visible. The recovery party could view the UAS data either on a Rover laptop or the UAS viewer in the case of small, Raven-like UAS. Alternatively, the UAS operator could simply pass the coordinates to the recovery party via voice radio.

Operational Feasibility

The concept is operationally feasible. The commander looking for the cargo would probably request assistance from a locally available reconnaissance, surveillance, and target acquisition (RSTA) squadron or aviation unit, or perhaps from a co-located maneuver headquarters. A RSTA squadron might provide either helicopter or UAS support. Circumstances will probably dictate whether the manned or unmanned aircraft is preferable.

Cost

The UAS option might be cost-free if the UAS is provided and operated by the owning maneuver unit. The logistics unit would be given the location information for the missing cargo via its organic radios. A 3G active RFID tag affixed to the 463L or similar platform might be a faster, more direct way (e.g., no dependence on cooperation from a unit with UAS or helicopters) to establish the cargo's location. Relying on the tag rather than tasking a UAS, the Army could avoid the opportunity costs associated with flying the UAS on the recovery operation.

Retrograde of Critical Items

As conceived here, a UAS would fly out to the point of failure, load the deadlined item aboard, and fly it to the appropriate maintenance echelon for repair. Value analysis of retrograde operations indicates that this concept is not cost-effective because it does not deliver a true benefit.[21] Movement of the faulty item from the point of failure to maintenance affects only a fraction of the retrograde chain, and the time saved in doing so is marginal—so marginal, in fact, that it would have no impact on the size of the stock of spares and inventory investments necessary to manage the uncertainties and delays associated with retrograde operations.

---

[21] Diener et al. (2004) and Wang et al. (2006).

CHAPTER FOUR

Acquiring and Employing UAS for Reconnaissance and Surveillance Applications

The previous chapter indicated which major logistics applications of UAS were feasible and infeasible, currently and in the near future. It noted that UAS-based reconnaissance and surveillance for logistics purposes, when supported by the capabilities of available sensors, would require large numbers of UAS. This chapter puts UAS-based applications into context with other options for securing linear assets (MSRs, pipelines) and area-point assets (LSAs, depots, fuel farms).

UAS May Contribute to Improved Logistics Situational Awareness

The old logistics security conception of operating from a relatively secure rear area and confronting only modest threats has been overtaken by events. Logistics units face the same threats as combat units. The common theme connecting most of the UAS applications—certainly all of the reconnaissance and surveillance ones—is a desire for better situational awareness to prepare logistics units to perform their tasks safely and effectively. After action reports, significant activity reports captured in the CIDNE and FusionNet databases, and interviews with logisticians returned from Iraq bear this point out. Today, logistics units, like their combat arms counterparts, depend on good situational awareness for their security and to help them accomplish their missions. Some areas and routes are safer than others, but in today's operations, without access to the network and all the sources of situational awareness–enhancing information on it, being outside the wire is just as risky for logistics troops as for anyone else. So long as present circumstances endure, logisticians will need the same situational awareness as the maneuver formations, requiring the network and communications tools to deliver it. At present, as Table 4.1 indicates, logistics formations do not have the density of radios, computers, and situational awareness tools that typical combat formations do.[1] Their situational awareness likely suffers as a result. Table 4.1 reflects the fractional amount of various situational awareness–producing items present in the unit property books relative to the types of units deployed in Iraq, Kuwait, and Afghanistan. Thus, the larger the number in each cell, the denser the item within the MTO&E for that type of unit. Consider SINCGARS radios as an example. Maneuver battalions have them in Iraq at a density of 3.9222, compared to the brigade support battalion or forward support battalion's density of 2.7771. The BSB/FSB has only 70 percent as many as the maneuver formation. The combat sustainment support battalions are likewise underprovided for at a density of 2.3074, or slightly less than 60 percent as many as found in maneuver battalions.

Figure 4.1 displays the information graphically. As the figure indicates, transportation battalions reflect the biggest shortfall in situational awareness–producing systems relative to their maneuver counterparts, but other logistics units also reveal lower densities of this equipment. Among the assets supporting situational awareness, the gap for UAS is among the largest.

As a first order of business, therefore, the logistics community should make the case for acquiring not only UAS but also radios, terminals, computers, and servers in the numbers necessary to support robust situational awareness across its forces. Because situational awareness is often a product of data from radars, airborne electro-optical/infrared (EO/IR) sensors, and fixed cameras, in addition to the other intelligence-collection disciplines, the community should examine its options for pulling all of these data sources down onto laptops. The

---

[1]   Based upon ARCENT G-4 data for task-organized units.

Table 4.1
Relative Density of Situational Awareness–Producing Equipment

	Iraq				Kuwait	Afghanistan			
	Maneu-ver	BSB/FSB	CSSB	Trans Bn	Maneu-ver	BSB/FSB	CSSB	Trans Bn
Number in data	30	9	12	2	9	2	2	1
Number of companies	6	6	5	8	5	8	5	1
Vehicles	196	206	142	361	109	197	95	68
SINCGARS	3.9222	2.7771	2.3074	1.3731	2.7200	1.4046	4.9421	3.0000
Handheld SINCGARS	1.4919	0.6400	0.3251	0.8669	2.8126	0.7964	1.1211	0.4412
SATCOM	0.0327	0.0243	0.0878	0.0028	0.7811	0.2366	0.3421	0.0000
HF	0.0146	0.0076	0.0100	0.0000	0.0458	0.0000	0.0000	0.0000
Handheld	0.9490	0.6071	0.4022	0.5936	2.3982	0.7786	1.1316	0.0000
STE	0.0078	0.0097	0.0053	0.0111	0.295	0.0102	0.0316	0.0000
Navigation sets	1.9687	1.6611	1.2845	1.2247	2.4043	1.8728	1.7421	0.2206
Convoy protection	0.7892	0.3551	0.3227	0.8821	0.3147	0.0662	0.2211	0.1029
Satellite phone	0.0206	0.0081	0.0241	0.0527	0.0316	0.0153	0.0000	0.0000
FBCB2	0.3236	0.1700	0.0065	0.0000	0.1069	0.0483	0.0105	0.0000
UAV	0.0184	0.0032	0.0006	0.0000	0.0356	0.0025	0.0000	0.0000
Driver's vision enhancers	0.1538	0.1479	0.0742	0.0097	0.0397	0.0000	0.0474	0.0000
Range finders	0.0667	0.0070	0.0000	0.0000	0.2006	0.0000	0.0211	0.0000
Night vision	3.8908	3.2909	2.6796	1.4022	5.9267	3.6056	4.4474	2.1618
Thermal sites	1.2147	0.4663	0.2939	0.0250	2.2475	1.1628	1.0579	0.0000

near-term objectives might be to train and equip all area/point assets (depots, LSAs, fuel dumps, etc.) with the means to deploy an integrated BDOC with capabilities for enhanced situational awareness and defense consistent with the threat and for linear assets (convoys, roads, and pipelines) to gain access to enhanced Sheriff's Net capabilities, including imagery products and predictive intelligence products (e.g., tools like Crystal and Multi-sensor Aerospace-ground

Figure 4.1
Relative Density of Situational Awareness–Producing Equipment

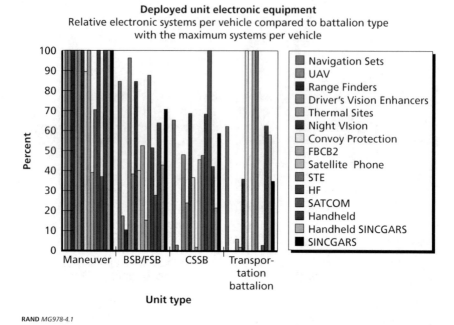

Joint Intelligence, Surveillance, Reconnaissance Interoperability Coalition (MAJIIC)).

Security Options for the Logistics Community

Once the logistics community acquires UAS and other assets needed to create connectivity to support enhanced situational awareness and to exploit additional data coming from a broader constellation of intelligence and security sources, it will want to have a clear notion of when it is appropriate to deploy assets from the menu of security resources in order to improve the security and survivability of its linear assets (convoys, routes, and pipelines) and its area/point assets (LSAs, depots, fuel dumps, etc.). This section of the chapter addresses these issues and highlights the important niche roles that UAS play in providing situational awareness.

Progressive Convoy and Route Security Steps

The Army's options for protecting its logistics assets lie in a space defined by the probability (or frequency) of attack against its routes and convoys on one axis and the cost to the United States of the damage associated with those attacks on the other. Figure 4.2 illustrates.

In this example, the probability or frequency of attack is represented on the vertical axis, and the resulting costs on the horizontal axis.[2] The gray line running through the middle of the chart represents increasing risk. Overlaid upon it are the progressively more effective options for protecting convoys and routes. They vary considerably in cost. UAS are a relatively high-cost option for convoy protection, but they provide an important level of capability to avoid having to abandon a route. In many theaters, alternative ground routes may not exist due to limited road systems. The figure indicates that the first course of action, as the probability of attack and cost of attacks begin to increase, is to harden the convoys that must move along the routes in question.[3] Hardening usually involves adding armor and security forces. The next option is to place fixed cameras to provide persistent coverage of dangerous areas. Following that option, the next step would be to build intervisible strong points able to exert positive control over the entire route. The UAS-based option appears right before the abandon route option. One might imagine these steps being applied as successive layers of defenses for convoys and routes. Layered defenses are typically preferred, because multiple layers mitigate against the consequences of leaks in an individual layer.

---

[2] The project team has calculated actual breakeven values based on data from CIDNE and FusionNet and presented them in a classified report.

[3] The specific order of the remedies may be theater-dependent, and some may not be practical in a given theater. For example, fixed cameras to cover the entire length of Afghanistan's ring road might prove impossible to maintain. Circumstances might dictate unique remedies not in the chart: a combat air patrol over a particularly vulnerable point or deployment of a JLENS (Joint Attack Cruise Missile Elevated Netted Sensor) or LEMV to maintain persistent surveillance. The point is that UAS tend to fit into a continuum of threat and cost.

Figure 4.2
Progressive Options for Defending Routes and Convoys

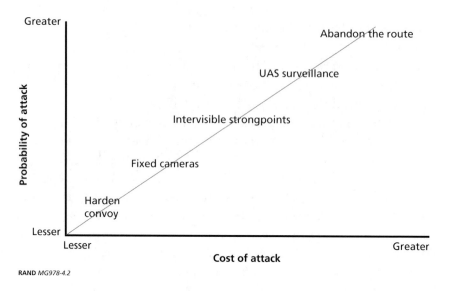

RAND *MG978-4.2*

Progressive Pipeline Security Steps

Figure 4.3 illustrates a similar set of options for protecting expeditionary pipelines. Ground monitoring is appropriate when the risks are low. Quartermaster troops in Iraq have patrolled their pipelines effectively in high mobility multipurpose wheeled vehicles (HMMWVs). If threats to the ground patrols escalate to unacceptable levels or unacceptable levels of damage to the pipeline occur despite these patrols, then the next option is to deploy UAS to monitor the pipeline. If the frequency and severity of attacks exceeds UAS ability to detect and counter them, the next option is to build pipeline exclusion zones, as the U.S. Air Force civil engineers have done with stretches of vulnerable pipeline in northern Iraq. If the costs of the attacks remain unacceptable, the next option is to install fixed cameras at all of the pumping stations for improved ability to monitor pipeline security and functionality. The ultimate step, if unacceptable levels of attacks or unacceptable costs continue, would be to construct intervisible strong points with fixed cameras, with QRFs being made available to secure the entire pipeline.

Figure 4.3
Pipeline Protection Options

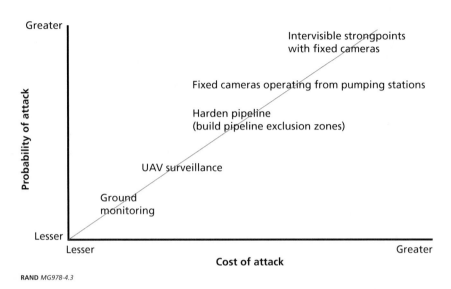

Intervisible strongpoints
with fixed cameras

Fixed cameras operating from pumping stations

Harden pipeline
(build pipeline exclusion zones)

UAV surveillance

Ground
monitoring

Probability of attack — Greater / Lesser

Cost of attack — Lesser / Greater

RAND *MG978-4.3*

Thus far, the more extreme measures have not been necessary to protect Army pipelines, but they could be needed in future operations.

Progressive Steps to Secure Area/Point Assets

A similar set of options exists for protecting LSAs, depots, fuel dumps, and similar logistics facilities. Figure 4.4 illustrates them. As the figure indicates, the first option is to establish a fixed perimeter for the installation, harden it against intrusion, and distribute the Army assets at the location to reduce their vulnerability to enemy indirect fire attacks (rockets, mortars, artillery).

If the frequency and cost of attacks remains unacceptable, the next step is to deploy counterfire radars and perimeter and tower cameras to increase the probability of detecting enemy activity. In some instances, aerostats may be desirable because they can provide multi-sensor payloads for persistent surveillance of the base and its surroundings. Some of these sensors can provide fidelity suitable to support targeting decisions.

Figure 4.4
Area/Point Asset Protection Options

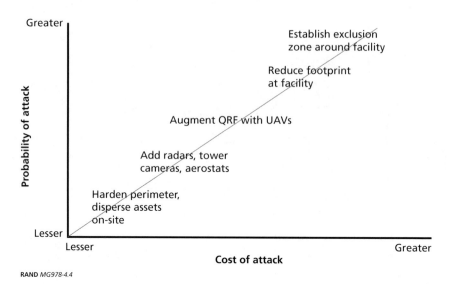

The next step in countering unacceptable frequencies of attack and costs of the consequences is to augment the QRF with UAS.

As noted in Chapter One of this report, QRFs supported by UAS in the study of LSA Anaconda/Balad Airbase enjoyed improved success rates in discovering and tracking perpetrators of indirect fire attacks against that installation.

If the frequency of attacks and their consequences still remain unacceptable, the next course of action is to reduce the logistics manpower at the facility. Doing so reduces soldiers' exposure to the threat, although such a course of action is not without consequences; a reduced presence may have operational impacts.

Assuming the facility remains vital to operations and further personnel reductions are impractical, the next step to improve security is to create an exclusion zone around the facility of sufficient depth to prevent future indirect fire attacks from short-range weapons and to force the enemy to rely on longer-range weapons in order to attack the installation.

None of the discussion presented in this section of the chapter is meant to suggest that a succession of steps like those described can deliver perfect security. Commanders will always find themselves managing risk during combat logistics operations. That said, logistics commanders could employ a succession of protective measures—some of them involving UAS—that could help them secure their convoys, routes, pipelines, and installations.

CHAPTER FIVE

Observations and Conclusions

UAS today are feasible solutions for enhancing several of the logistics operations examined in this monograph. For some operations, such as route surveillance and installation security, large numbers of UAS would be required. For this reason, the advent of smaller, less expensive UAS à la Scan Eagle and its peers is a welcome development. Currently, UAS would make their greatest contribution in supporting QRF responses to indirect fire attacks. In addition, there may be circumstances such as mountainous terrain where UAS overcome line-of-sight limitations of radars and fixed cameras, giving UAS a comparative advantage over options that they do not enjoy in other conditions, such as those often found in Iraq. UAS offer an option to help locate off–drop zone cargos. Using them for predeployment theater reconnaissance is feasible if OPSEC and benefits issues are addressed. At current prices and sensor capabilities, UAS are not the most cost-effective option for all logistics operations for which they are feasible. But their relative attractiveness can be expected to increase, perhaps rapidly, as system costs fall and sensor capabilities improve.

Figures 5.1, 5.2, and 5.3 present our estimates of the net utility of UAS for logistics applications, today, in the near term, and in the longer-term future, respectively. Figure 5.1, reflecting current technology, follows the earlier analysis in this monograph that found the use of LIDAR to ascertain river navigability to be technically infeasible. Theater reconnaissance, while technically feasible, failed the benefits test when compared to currently available operations support materials. Retrograde of critical unserviceable items also failed the benefits test.

Figure 5.1
UAS Utility for Logistics Applications in 2010

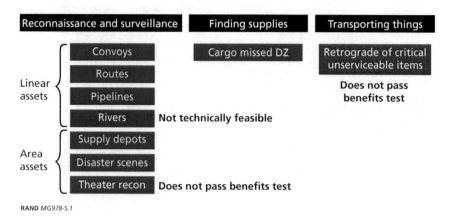

RAND *MG978-5.1*

Figure 5.2 illustrates the research team's estimate of UAS utility for these tasks in the near-term future. The figure reflects the team's judgment that in the near term, UAS suitable for theater reconnaissance could be stealthy, overcoming the OPSEC concern, and could carry payloads optimized for logistics-oriented terrain studies and similar tasks.

Figure 5.2
UAS Utility for Logistics Applications in the Near Term

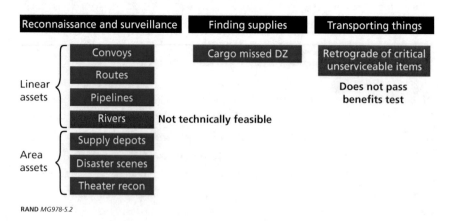

RAND *MG978-5.2*

Figure 5.3
UAS Utility for Logistics Applications in the Long Term

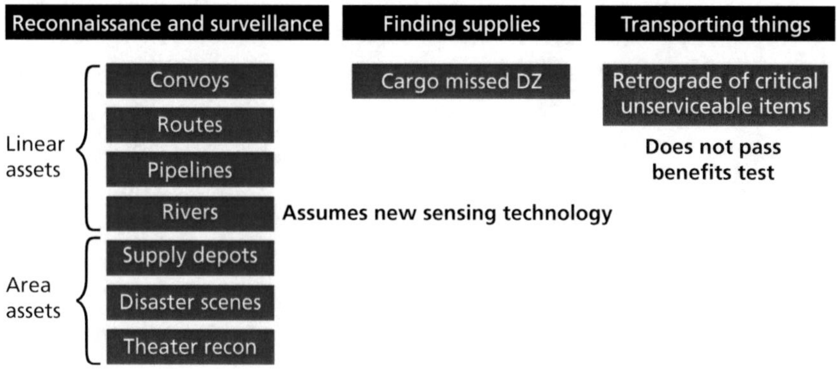

Reconnaissance and surveillance	Finding supplies	Transporting things

Linear assets:
- Convoys
- Routes
- Pipelines
- Rivers

Cargo missed DZ

Retrograde of critical unserviceable items

Does not pass benefits test

Assumes new sensing technology

Area assets:
- Supply depots
- Disaster scenes
- Theater recon

RAND *MG978-5.3*

Figure 5.3 presents the team's judgments about the long-term future and, specifically, the expectation that in the ensuing 14 years, a new sensing technology may emerge that is suitable for determining river navigability.

The widespread interest in UAS for logistics activities highlights a broader issue confronting logisticians today: acquiring and maintaining sufficient situational awareness to operate safely off of FOBs and along often-dangerous MSRs. As shown in Chapter Three, logistics units do not enjoy the same density of C4I assets as combat arms units, despite the fact that in today's operating conditions, they require access to the same information that combat arms units use to develop their understanding of their circumstances: where the enemy is, where other friendly forces are, etc. Our interviews lead us to believe that the logistics community may be able to take better advantage of currently available resources that could improve its security and situational awareness that combat formations routinely exploit.

If our interview-based impressions are correct, then the Army should make it a priority to provide additional C4I equipment to logistics units; this equipment includes UAS but also much more. Logistics units require the same situational awareness as maneuver formations in order to operate safely and effectively. In addition to UAS, logisticians therefore would benefit from additional computers, servers, and radios that make maximum situational awareness possible. A full DOTMLPF (doctrine, organization, training, materiel, leadership and education, personnel, and facilities) analysis of how best to enhance the situational awareness of logistic units appears warranted.

If the Army decides to acquire more UAS for logistics units, it should be clear about its objectives in this regard. It seeks situational awareness to allow its logistics units to detect and avoid threats where possible, and capabilities to destroy those threats under circumstances consistent with the rules of engagement and the primary missions these units are tasked to accomplish.

Detailed Analysis of Selected UAS Logistics Concepts

This appendix provides more detailed analyses of the UAS logistics applications that fall generally under the heading of reconnaissance and surveillance activities. It addresses each of the potential logistics applications in the order in which they appear in Figure 1.1, beginning with convoy overwatch.

Convoy Overwatch

Overwatch is a maneuver technique in which one unit provides surveillance along the route of advance while another unit moves ahead along that route. The overwatching unit must be in a position to observe the route and in range to provide fire support to the advancing unit if it encounters the enemy.[1] If the two units at some point exchange roles—the lead maneuver unit halts and provides overwatch as its sister unit moves ahead along the line of advance—the two units are said to be practicing bounding overwatch.

UAS-based convoy overwatch, as conceived in the logistics community, envisions an armed UAS performing a modified type of bounding overwatch with a logistics convoy. The UAS coordinates its movement with that of the convoy, searching the road ahead for signs of trouble: an IED, an ambush, a sniper, etc. As the convoy traverses

---

[1] See Headquarters, Department of the Army (2002), Chapter 3, for a more detailed description of overwatch.

47

one segment of the route safely, the UAS flies ahead to the next series of hot spots (sites of previous attacks or terrain features that make attacks there attractive). The UAS data, FMV, and real-time SAR can be monitored by the convoy commander and by a fixed headquarters, perhaps a movement control center. The convoy commander talks to the UAS operator to get the system to look at certain sites or to conduct reconnaissance by fire or perform other tasks along the route. When the UAS spots potential trouble, the convoy commander can take several courses of action: conduct a reconnaissance by fire to "trip" the ambush or detonate the suspected IED; use the UAS to suppress the suspect site, allowing the convoy to drive through as quickly as possible; use the UAS to scout an alternate route around the scene; or halt, assume hasty defensive positions, and use the UAS for local security until help arrives. Because convoy overwatch would find and disrupt enemy attacks before they could engage logistics convoys, using UAS in convoy overwatch would reduce the probability of attack on such convoys and, when attacks occur, reduce the amount of damage they cause, because the convoy would take evasive action rather than simply drive in to the kill zone of the attack.

Case Zero

Interviews with logisticians who conducted operations in Iraq in 2008 shaped our understanding of convoy security practices there. There are two qualitatively different types of logistics convoys on the roads currently in Iraq: those that bring supplies up MSR Tampa from Kuwait to the FOBs, and those that subsequently push supplies forward from the FOBs to their subordinate or dependent outposts. In the case of the former, 19–23 convoys cross the frontier from Kuwait daily and drive up MSR Tampa to resupply the FOBs. These convoys typically contain 40-some vehicles, with one security vehicle for every ten logistics vehicles. The convoys have infantry embarked with them to provide security. When they encounter a coordinated attack, they typically "push through" and continue their mission. Rerouting is unusual. Interviewees indicated thst coordinated attacks (as opposed to harassing actions like random small arms fire) occurred "every two weeks" but that damage was limited typically to fragmentation.[2]

A second set of interviews involved officers with earlier experience from 2003 to 2004. These officers indicated that during these early OIF tours the IED threat was "immature." They conducted 24-hour-per-day operations, initially from Camp New York in Kuwait, then later splitting operations from FOB Speicher out to its dependent outposts and from Camp Taji out to its supported outposts. Their convoys were smaller, on the order of 20 cargo vehicles plus security vehicles (one gun truck per six cargo carriers). One in ten convoys suffered attacks initially, but some routes sustained an attack "every other night." They experienced well-aimed sniper fire, IEDs, explosively formed penetrators (EFPs), and harassing attacks involving grenades and small arms fire. The interviewees noted that sometimes maneuver convoys had UAS support, and that sometimes the Marines would escort their convoys with AH-1 armed helicopters, but they were untrained in how to use these assets and would not have been able to employ the AH-1s effectively in an attack. They also noted that the damage their convoys suffered was light.[3]

Technical Feasibility

Technical feasibility for current-generation UAS in the role of convoy overwatch can be determined through the answers to five questions:

- Are there UAS suitable for the mission as imagined?
- Are there sensors that could detect the threats logistics convoys seem likely to confront?
- Are there data links to deliver the sensor data to the convoy?
- Is there ordnance that UAS could deliver in an overwatch role?
- Do UAS currently have the endurance to provide overwatch?

The existence of armed UAS like Predator and Reaper suggest the answer to the first question is "yes." The Army Sky Warrior could also

---

[2] Interviews with officers from the 311th Expeditionary Sustainment Command, March 30 and June 5, 2009.

[3] Interviews with officers from the 377th Transportation Battalion, April 2, 2009.

be a candidate. Indeed, a number of new UAS are debuting that might be useful in this role: Yellow Jacket to find bomb-planters, Copperhead with SAR, Sentinel Hawk for route surveillance, and the Autonomous Rotorcraft Sniper System (ARSS).[4]

Today's sensors, specifically FMV and SAR, suggest that sensing capabilities are technically feasible, although at present there are limitations with these technologies that have to do with the quality of sensor data and rules of engagement discussed below under operational feasibility.

Data links to get the data off the sensor and onto the laptops of a convoy commander also exist. Rover is the archetypical example, but the Raven tactical vehicle's viewer suggests still simpler approaches.

Armament options are abundant, and therefore feasible. Predator and Reaper carry Hellfire missiles. Hydra 70 rockets offer at least nine different warhead options that might make them an attractive option where rules of engagement are restrictive. The small diameter bomb, automated sniper rifle, and eventually the Spike missile will also be available.

Long endurance is feasible. Army Sky Warrior is capable of 30 hours aloft.[5]

Operational Feasibility

Operational feasibility for convoy overwatch is contingent upon the UAS's probability of detection given the threats to the convoy and the chain of command's responsiveness when a threat to the convoy is identified. Operational feasibility is therefore theater-dependent. Today's convoys face multiple threats, including sniper fire from prepared positions (often in buildings), random fire from clusters of buildings, grenades thrown from a crowd, rocket-propelled grenade (RPG) fire aimed at soft-skinned vehicles, and a host of IEDs. These IEDs include those that are command-detonated at choke points (bridges, overpasses, road

---

[4] See Kris Osborn, "Pentagon Readies 3 Anti-IED UAVs," *Navy Times*, December 16, 2008.

[5] See "Sky-Warrior ERMP UAV System," *Defense Update*, fact sheet. http://defense-update.com/products/w/warriorUAV.htm

intersections, tunnels); pressure-activated devices on the road in plastic bags or buried; manually thrown devices from vehicles traveling in the same direction as the convoy or dropped off of bridges, thrown from the side of the road, or employed by suicide bombers either as individual assailants or in vehicles (the so-called vehicle borne IED or VBIED).[6]

Inherent in the overwatch concept is the notion that the convoy commander or one of his subordinates will be able to understand the data coming off the UAS in terms of its threat potential; that is, he or she must be able to understand "in real time" (almost immediately) what the sensor is "seeing." This requirement for nearly instantaneous understanding limits the sensor choices to FMV and SAR, and perhaps SIGINT. Table A.1 suggests the rough suitability of likely sensors given the threat.

A green checkmark in a table cell indicates that the sensor in question may have some capability to detect the threat, and an empty

Table A.1
Suitability of Sensors to Detect Threats

Threat/Sensor	FMV	SAR	SIGINT
RPG fire	√	√	√
Random fire from buildings			
IED emplacement team	√	√	
Sniper from prepared position			√
Grenade thrown from crowd			
Command-detonated IED at choke point			√
Pressure-activated IED in/on road		√	
Thrown IED			
Suicide IED			

NOTE: Red shading indicates most dangerous threats to logistics convoys. Orange shading indicates second-most dangerous threats.

---

[6] List compiled from the 181st Transportation Battalion's unclassified after action report available on JWICS.

cell indicates the sensor has little or no capability with respect to the threat.[7] Snipers in prepared positions are very difficult to detect visually (i.e., with electro-optical sensors). There are acoustic detection systems, but they are not suitable in the context of the UAS overwatch concept.[8] SIGINT offers some possibility, but only if the sniper is actively communicating with his spotters or chain of command at the time.

None of the sensors in the table have any effectiveness against random small arms fire or grenades thrown from a crowd. In both types of attack, the enemy's presence is nearly undetectable until the event; prior, he is simply an anonymous figure indistinguishable from anyone else in the crowd. The assailant's dwell time at the scene of the attack also tends to be very short, perhaps measured in seconds as he moves to curbside, throws the grenade, and flees down an alleyway, or alternatively, retrieves his rifle from its hiding place, takes a few poorly aimed shots, returns the rifle to its hidden place, and leaves to go about his normal business.

All three sensors may detect an RPG gunner, if only a few seconds before he fires. SIGINT might discover him if it receives an electronic go-ahead from his spotter or from his commander. FMV and SAR might detect him as he takes his firing position, which typically requires line-of-sight to the target at a range of less than 500 meters, but often much closer distances (about 100–200 meters). RPG gunners usually employ fire-and-flee tactics, dashing into nearby buildings where they are protected from return fire by the presence of innocent civilians.

IED emplacement teams are vulnerable to detection by FMV and SAR. Although their work does not usually require them to linger at the emplacement site for an inordinate amount of time, emplacing an IED does require a longer dwell time than throwing a grenade or engaging in harassing small arms fire. Indeed, Task Force (TF) ODIN

---

[7]   A National Imagery Interpretability Rating Scheme (NIIRS) score of 6 to differentiate between armed and unarmed personnel.

[8]   The acoustic sensors have to be mounted on the ground so that their geometry remains constant, allowing triangulation of the sniper's position through the measurement of time delays for the sound of the rifle shot to arrive at each sensor.

has taken advantage of this vulnerability in its offensive counter-IED operations.[9]

Command-detonated IEDs at choke points are difficult to detect because they do not offer clear identification criteria in areas already littered with trash. SIGINT may offer some capability to detect them if the combatant with the detonator is listening for reports from his spotters (telling him the convoy is near) or for instructions from his commanders. That said, any kind of basic communications discipline on the enemy's part may mean there are simply too few communications events to fix the bomber's location.

Pressure-activated IEDs present similar detection problems. Change detection, FMV, and SAR may have some utility against such attacks if the IED is embedded in the road surface, leaving residue from the excavation that the sensor could sense. If the bombers' dwell time grows longer, then FMV may become more effective against them.

Thrown IEDs and suicide IEDs suffer from the same detection shortcomings ascribed to grenade-throwers. The perpetrators rarely present a clearly identifiable signature indicating their intent until they commit the act, at which point it is too late; they have accomplished their mission.

Eight of the 27 cells in Table A.1 contain green checkmarks indicating some capability for detection. This means that the sensors employed in the UAS overwatch concept would be useful against about 30 percent of the threats facing logistics convoys. Against the most dangerous, red-coded threats, only 2 of the 12 cells—less than 20 percent—show an effective sensor. The next set of considerations in attempting to estimate operational feasibility is the responsiveness of the chain of command, often characterized as "kill chain responsiveness," highlighting the sequence of events that must be optimized in order to provide effective fire support as a part of the convoy UAS overwatch concept. Figure A.1 illustrates the key considerations in determining kill chain responsiveness.

---

[9] According to Semi-Annual TF ODIN Meeting, April 23–24, 2008. See also "Army Praises ODIN," *Aviation Week*, May 14, 2007.

Figure A.1
Factors Determining Kill Chain Responsiveness

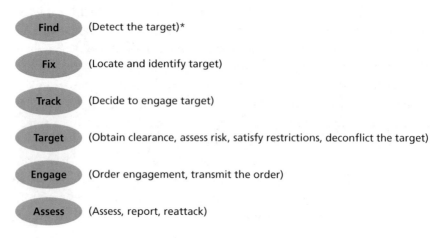

Find   (Detect the target)*

Fix   (Locate and identify target)

Track   (Decide to engage target)

Target   (Obtain clearance, assess risk, satisfy restrictions, deconflict the target)

Engage   (Order engagement, transmit the order)

Assess   (Assess, report, reattack)

*Adapted from Figure II-4, Dynamic Targeting, in Joint Pub 3-60, *Joint Targeting*, April 13, 2007.

RAND *MG978-A.1*

For the UAS overwatch concept to be operationally feasible, the convoy commander must be able to understand the system's sensor information in terms of what it says about the convoy's safety, formulate a course of action in response to this information, and execute the chosen course of action before the convoy reaches the point along the route where the UAS sensed a threat. Table A.2 illustrates the issue.

Interviewees for the project indicated that logistics convoys typically travel at 40 miles per hour.[10] Thus, if the convoy commander is to have 30 seconds to comprehend what the UAS sensors "see" and formulate and execute a course of action to address it, the UAS must be 536 meters ahead of the convoy. If the convoy commander is to have 60 seconds in which to work (e.g., target the suspects and fire the system's ordnance at them), the UAS must be 1,073 meters ahead of the convoy.

------

[10] Convoys in Afghanistan are much slower: typically they travel ~30 MPH on the ring road, 10–15 MPH on semi-improved roads, and ~5 MPH off road, according to information from the 801st BSB.

Table A.2
Time-Distance Issues with UAS Overwatch

Threat/Sensor	FMV	SAR	SIGINT
RPG fire	√	√	√
Random fire from buildings			
IED emplacement team	√	√	
Sniper from prepared position			√
Grenade thrown from crowd			
Command-detonated IED at choke point			√
Pressure-activated IED in/on road		√	
Thrown IED			
Suicide IED			

Having 90 seconds for the convoy commander would mean flying the UAS 1,609 meters ahead of the convoy. Alternatively, a convoy commander could slow his/her rate of march to maximize response time to UAS data, although doing so involves some risks. For example, a slower convoy could be more vulnerable to undetected snipers.

If each of the steps in Figure A.1 takes five seconds, then a UAS lead of 536 meters would be adequate. If each of the steps takes ten seconds, the process would take 60 seconds, which would require a 1,073-meter lead. If any of the steps takes longer (the actual targeting process of obtaining clearance, doing a risk assessment, satisfying local restrictions, and deconflicting the target, for example), the entire process could take multiple minutes, which would mean that the convoy would arrive at the site of potential trouble before the commander has been able to employ his/her overwatching UAS to deal with the threat.

Alternative tactics might improve the concept's utility in Afghanistan, where convoys typically move more slowly and their commanders therefore enjoy longer reaction times. In the alternative TTP, the UAS might focus on hot spots ahead of the convoy. The UAS could conduct reconnaissance by fire in the more remote areas without endangering population in attempts to detonate suspected IEDs. The concept would be sensitive to the quality of intelligence.

Route Surveillance[11]

Case One

Route surveillance differs from the previous convoy overwatch concept in several respects. First, the surveillance effort orients on the road rather than on a convoy moving along a road. Second, the mission is surveillance rather than overwatch, so the UAS doing the job need not be armed. As the concept has been envisioned, the UAS would attempt to detect suspicious activities as they occurred, and would thus be dependent upon sensors that could produce real-time results, typically FMV, SAR, and SIGINT.[12] If we imagine UAS patrolling 100-mile (160,934 meters) stretches of an MSR at 135 knots true air speed (KTAS) with perfect sensors and an enemy that never lingers longer than four minutes along the road, we can calculate the probability of detecting him as follows. Each side of the orbit is 38 minutes or 2,280 seconds long, so the UAS transits a 500-meter route segment every seven seconds. The enemy is present at any one location only for 240 seconds. In 240 seconds the UAS can fly about 16,940 meters. If an attack happens somewhere in the orbit not close to the turnaround points (again, assuming perfect sensors), the enemy can be detected as long as the UAS is within (16,940 + 250) meters behind the point of attack or 250 meters ahead of it. The probability of detection is:

$$\frac{2\,(16,940 + 250 + 250)}{160,934 \times 2} = {\sim}10.8 \text{ percent}$$

If the attack takes place at the turnaround points for the orbit, the probability of detection drops to approximately 5.4 percent, and if the

---

[11] This application of UAS has no single Case Zero. Typically MP and Engineer units provide route clearance and search MSRs for IEDs and similar threats. Maneuver units sometimes have route security responsibilities.

[12] Real-time capabilities are growing. DARPA, in one example, is launching a new effort dubbed "TAILWIND" (tactical aircraft to increase long wave infrared night-time detection), meant to provide "a medium-area persistent day/night surveillance capability with real-time image products to multiple users." Richard Scott, "DARPA takes advantage of TAILWIND for UAV surveillance," *International Defense Review*, May 20, 2009.

attack occurs within 8,720 meters of the turnaround points, the probability varies between 5.4 and 10.8 percent with the specific location of the attack.

Results in Afghanistan might be significantly different, however, given the enemy's preference for large IEDs buried in the road, and the longer periods of time needed to dig the hole, embed the IED, and cover it. If the enemy required 20 minutes, the UAS could fly 72,000 meters. Using the same formula as above, the probability of detection would be 100 percent.

Unfortunately, today's real-time sensors, as Table A.3 indicates, are able to detect only a few of the threats that stalk MSRs. As the table indicates, the sensors and SIGINT are only effective against about 30 percent of the threats, and SIGINT is contingent upon the enemy communicating with his spotters or chain of command. If he operates silently, SIGINT loses its value. Thus snipers and IED operators, if they do not communicate with spotters or their chain of command, will not be found via SIGINT. Pressure-activated IEDs may be detected by SAR, but only if the weapon is actually embedded in the roadway and there is displaced soil for the SAR to sense.

Table A.3
Threats and Sensors for Route Surveillance

Threat/Sensor	FMV	SAR	SIGINT
Sniper from prepared position			√
Random fire from buildings			
Grenade thrown from crowd			
RPG fire	√	√	√
IED emplacement team	√	√	
Command-detonated IED at choke point			√
Pressure-activated IED in/on road		√	
Thrown IED			
Suicide IED (including VBIED)			

SOURCE: Based upon the 181st Transportation Battalion after action report.

Case Two

We developed a second version given the limitations of the first. In the latter version, the UAS carries advanced sensors chosen for their capabilities in change detection and high-resolution imaging even though the outputs from these sensors require processing to render them useful, e.g., the Landeta BuckEye System.[13] The idea was that enhanced imaging could reduce the risk to units tasked with sweeping a route and make them more effective in finding IEDs and ambush sites.

Technical Feasibility

Technical feasibility for Case One is at best marginal with current sensor capabilities, given the small probability of detection (under 11 percent) and the limited utility of the sensors against the threats (able to detect only 30 percent of them).

Case Two is technically feasible, with qualifications. Hyper and multispectral imagers deliver high-quality images that often reveal evidence of movement and disturbance of terrain that could be valuable in identifying enemy infiltration routes, camouflaged fighting/firing positions, IED positions, and similar items of interest. The current generation of these sensors have high false alarm rates, the signatures they seek are fragile, and they may be confused by changes in enemy TTPs. The sensors themselves are sensitive and can require intensive maintenance.

Operational Feasibility

Operational feasibility for Case One is poor because of the limited dwell time of the enemy (less than four minutes) and typical QRF reaction times (30–90 minutes).[14] Even if the sensors detect the enemy, the logistics convoys moving along the road must continue their mission and deliver their cargo. Experience in both Afghanistan and Iraq

---

[13] Described in First Lieutenant Matthew D. Brady, "The 766th Explosive Hazards Coordination Cell Leads the Way into Afghanistan," *Engineer*, January–April 2009, pp. 42–47.

[14] Interviews with 311th Expeditionary Sustainment Command personnel, March 30 and June 5, 2009.

indicates that logistics convoys have few opportunities for detours because of the limitations of the road networks.

Operational feasibility for Case Two is contingent upon the TTPs adopted by the convoys and units that own the terrain the convoys transit, and upon the C4I architecture to get the data from the UAS through processing and annotation and out to the convoys and security units on the MSR. If the UAS missions can be synchronized with convoy schedules so the imagery is available in time for the convoys and local security units to benefit from it, and if the C4I architecture will support delivering the imagery on a just-in-time basis, then the concept is operationally feasible.

Some live experimentation would be necessary to perfect the TTP.

Area Assets

Logistics sites could include supply caches, depots, fuel bladder farms, retrograde yards, and perhaps others. The basic concept for the employment of UAS would place an orbit above the logistics site with a FMV sensor on board to provide real-time surveillance of the perimeter and to warn of people approaching or launching indirect fire attacks aimed at the installation. When the UAS senses the approach of unidentified personnel or activities consistent with an indirect fire attack, the site commander launches the quick reaction force to investigate. The concept also includes using the UAS to track unidentified personnel in the aftermath of an attack.

Case Zero

This concept is feasible and cost-effective; UAS's comparative advantage over fixed sensors is in tracking attackers. Today's fixed surveillance practices, at least at larger installations, already surpass those in the UAS employment concept.[15] (See Figure A.2.) First, fixed cameras including rapid aerostat initial deployment (RAID) towers and wide-

---

[15] The practices described are based upon an extensive examination of Balad Airbase/LSA Anaconda in Chow et al. (2009).

Figure A.2
Balad Airbase/LSA Anaconda Integrated Defenses

RAND *MG978-A.2*

area infrared surveillance thermal imager (WISTI) cameras along the
perimeter provide a multiple-kilometers-deep view of the surrounding
countryside and are integrated into the base defense operations center
along with omnidirectional radar, countermortar radar, and sometimes ground surveillance radar systems. All of the surveillance systems are further integrated with the Air Force TASS warning system,
which sounds an alarm based on the radar-generated predicted point of
impact for incoming mortar, rocket, and artillery fire.

In the aftermath of an attack, UAS have proved more effective
than other options at finding the fleeing attackers and guiding a QRF
to arrest them. Unfortunately, the enemy needs only seconds to launch

an indirect fire attack, so when the enemy is detected, the attack is already under way; the United States can only hope to find and track the assailants and arrest them. It is worth noting that even if suspicious people are sighted in proximity to a U.S. installation, preemptively firing on them is bad policy in counterinsurgency operations, where wounding or killing civilians can undermine whatever fragile success has been achieved to date. QRFs without some type of support have often fallen prey to ambushes as they tried to catch up to the enemy gunners. When QRFs have had helicopter support to track the assailants, they have never found them.

Case One: A UAS Orbit Above a Logistics Site

In this instance, an Army logistics unit has established some kind of site and seeks to improve its security by placing a UAS orbit overhead so the site command post is aware of people moving about in proximity to the site perimeter and of attempts at intrusion. The logistics site itself is small enough to allow complete sensor coverage with a single orbit.

Technical Feasibility

Technical feasibility of current UAS rests on answers to most of the questions posed in the convoy overwatch analysis:

- Are there UAS suitable for the mission as imagined?
- Are there sensors that could detect the threats likely to confront logistics sites?
- Are there data links to deliver the sensor data to the logistics site?
- Do UAS currently have the endurance to provide long-duration coverage of the site in question?

Logistics site surveillance via UAS is technically feasible. There are several models of UAS, including Hunter and Sky Warrior, that could perform the basic mission (though it would take more aircraft to sustain the orbit using Hunter UAS rather than Sky Warrior). There are sensors capable of National Imagery Interpretability Rating Scale (NIIRS) 7 that might offer resolution sufficient under optimal conditions to discriminate between an armed man and a man with a shovel.

Although typical electro-optical sensors flying on today's UAS at typical operating altitudes do not yet achieve a score of 7, the current pace of miniaturization and refinement of sensors suggests that NIIRS 7 may be more widely available in mission packages in the near term or with today's sensors operating at lower altitudes, with longer focal lengths, or using frame sampling. The data links and laptops for transferring and monitoring the data currently exist, although there may be limitations on their ability to preserve the level of resolution present at the sensor (e.g., Rover III, the third-generation laptop monitor), and 30 hours endurance has been delivered.

Operational Feasibility

For the concept of UAS site surveillance to be operationally feasible, it must be able to detect a threat to the site and do something useful in response to it. Table A.4 summarizes typical threats.[16]

UAS surveillance of a logistics site is operationally feasible, but its utility and cost-effectiveness depend upon local conditions. The sensors of today cannot detect a significant percent of the threats confronting such facilities and fare poorly at establishing hostile intent for several others, including indirect fire attacks and VBIEDs. For example, in Iraq, UAS are not very suitable for detecting snipers firing from well-

Table A.4
Threats to Fixed Sites

Threat/Sensor	FMV	SAR
Sniper in prepared position		
Random small arms fire		
Preparation for indirect fire attack	√	√
Indirect fire attack		
Perimeter breach	√	√
VBIED attack	√	√

---

[16] Chow et al. (2009).

prepared positions or for detecting harassing fire coming from a building or cluster of buildings.

By contrast, utility for the threat of preparation for indirect fire attack is higher for the Iraq case. Assuming the sensor is capable of NIIRS 6, the UAS might alert the sensor monitor to suspicious activity consistent with preparations for an indirect fire attack if there is other supporting information, perhaps a tip from a HUMINT or SIGINT source. The enemy gunners often use the same launch tube over and over, and leave it camouflaged at the firing site. Thus, while the sensor may detect men at the scene, it will not detect many clues establishing hostile intent until the mortar or rocket fires. FMV and SAR should, however, detect anyone attempting to approach the perimeter close enough to penetrate it. The issue will be the response—whether the manpower on-site can mount an effective counterattack. Likewise, the sensors in question would be able to detect a vehicle approaching the entry control point (or any part of the perimeter for that matter), but they will not detect the explosive device on board, which creates an opportunity for a successful attack.

Moreover, the concept does not integrate the UAS with other sensors (cameras and radars) or with other intelligence-collection disciplines (HUMINT, SIGINT) that have proven to be critical in counterinsurgency operations.[17] Finally, the concept simply superimposes the UAS on the logistics site without incorporating any of the recent developments from C-RAM, the U.S. Army's Counter-Rocket, Artillery, and Mortar initiative, which has employed terminal defenses to destroy incoming enemy fire before it can detonate and cause injuries and damage.

UAS in Support of Disaster Relief

The second area-oriented application for UAS involves a disaster scene with mass casualties, and the potential for significant contamination: radiological, chemical, or biological. In this concept, UAS perform an aerial damage assessment to assist authorities in gauging the scope of the disaster scene, the direction and extent of any downwind plume,

---

[17] Peters et al. (2010).

the nature of any contamination, and differentiation within the site between areas where survivors might be found and areas where the dead must be recovered. As they did during the aftermath of Hurricane Katrina, they can also serve as communications relays.

As originally conceived by the logisticians who nominated this concept, the UAS would be instrumental in moving human remains as a part of recovery operations. In true mass casualty events, however, the Department of Health and Human Services suspends the practice of recovering remains, decontaminating them, and preparing them for individual burial, and instead employs mass graves.[18] Moreover, every effort is made to keep contaminated items inside the hot zone and to limit transfers of material between the hot zone and the outside, thus limiting the spread of contamination. Unmanned vehicles may still play a role in handling human remains, but in all likelihood they would be unmanned ground vehicles (UGVs), because they already exist in explosive ordnance disposal (EOD) units and as part of the Army's future force plans. UGVs have survived the Future Combat System's cancellation of the manned vehicle programs.

Case Zero

If such an event were to occur today, local emergency agencies and law enforcement would be the first responders.[19] Mutual aid compacts with neighboring jurisdictions would be likely sources of additional capacity. As local authorities begin to appreciate the scope of the disaster, they ask for assistance, first from their state government; subsequently, if he or she deems it necessary, the governor requests assistance from the federal government. An official declaration of emergency makes many assets available, including Army and Air Force aircraft, to survey the extent of the damage and to ascertain whether any contaminants are present. National Guard units assist local law enforcement in establish-

---

[18] See standards at Health Systems Research, Inc., *Altered Standards of Care in Mass Casualty Events*, Rockville, Md.: Agency for Healthcare Research and Quality, U.S. Department of Health and Human Services, AHRQ Publication No. 05-0043, April 2005.

[19] For a detailed description, see Eric V. Larson and John E. Peters, *Preparing the U.S. Army for Homeland Security: Concepts, Issues, and Options,* Santa Monica, Calif.: RAND Corporation, MR-1251-A, 2001.

ing a site perimeter and might employ nuclear, biological, and chemical (NBC) reconnaissance vehicles organic to their formations or perhaps use radiac meters and chemical swipes from their field NBC kits to help detect the outer edges of the contaminated zone and to determine what sort of contamination is present. Engineer units—active and reserve components from all services—and local construction companies would support the on-scene commander by clearing pathways into the disaster site to enable rescue operations. If radiological contamination is present, officials will have to determine dose-rate policy for rescue workers. Search and rescue efforts could employ helicopters and fixed-wing aircraft to survey the site looking for signs of life.

Recovery operations begin when officials conclude they have rescued everyone left alive. Recovery parties with cadaver dogs search for the dead. Suitable mass gravesites are identified and excavated within the hot zone. Human remains are identified if possible, then interred in the mass grave. Recovery workers use equipment such as UGVs, forklifts, and bulldozers to limit their exposure to the contaminated remains, and otherwise take protective actions as prescribed by public health officials.

Technical Feasibility
Technical feasibility for current UAS rests on answers to questions similar to those posed in the earlier cases:

- Are there UAS suitable for the mission as imagined?
- Are there sensors that could detect the contaminants and establish the extent of the contamination?
- Are there data links to deliver the sensor data to the on-scene commander?
- Do UAS currently have the endurance to provide long-duration coverage of the site in question?

Suitable UAS exist today. Sky Warrior and the other Predator derivatives have the payload capacity to handle the sensors that would be required. The sensors exist, including sampling technologies like those employed by WC-130 aircraft and spectroscopy sensors like those employed in commercial industry.[20] Data links and UAS endurance have been established in the earlier cases.

Operational Feasibility

As noted in Chapter Three, Congress has instructed the Department of Defense, Department of Transportation, and the Department of Homeland Security to find a solution. NORTHCOM and the Army's stationing plan currently call for the distribution of UAS across the United States so that they will be readily accessible to support disaster relief operations.

Theater Reconnaissance Prior to Deployment

This concept anticipates using Global Hawk or similar high-altitude, long-endurance UAS to support logistics planning for a new theater of operations. Such a UAS mission need not be exclusively in support of logistics planning; a single sortie could easily collect data in support of many consumers. The logisticians would use the resulting information to supplement their force laydown planning, identifying trafficable routes, attractive areas for various logistics facilities, and identifying mobility constraints.

Case Zero

Currently, a number of products could be used to support predeployment planning like that described in the paragraph above. The series of Defense Contingency Products includes route studies, operational support studies, and Joint Theater Transportation Studies. In addition to these resources, logistics planners have access to the NGIC GIS data and satellite imagery, and to USTRANSCOM's SIPRNET site, which includes a "transportation intelligence" area with the latest intelligence relevant to transportation and logistics operations. Planners can also

---

[20] See, for example, the gas vapor monitoring technology at "What's New at Brimrose! SMART PROCurement 2009 Trade Show," *Brimrose*, web page.

take advantage of commercially available satellite imagery from sources like Digital Globe, Google Earth, and Geosage, among others.

Case Two

Instead of flying UAS for predeployment information on conditions in a theater of interest, the Combatant Command (COCOM) could produce a formal intelligence production requirement and call on the NGIC to satisfy it. NGIC would draw on all intelligence sources, including national technical means, and produce a comprehensive study to answer logisticians' concerns about the theater. Ideally, the logistics community should levy intelligence production requirements in support of all the contingency plans logisticians might have to execute.

Technical Feasibility

Using a UAS to collect relevant information to support predeployment theater reconnaissance is technically feasible. Current systems can perform this task with a variety of payloads.

Operational Feasibility

We assess the use of UAS for this task to be operationally feasible, although we are uncertain about the marginal additional value of the information gathered by the UAS relative to intelligence and data from the sources noted in Case Zero and Case Two, above, and there are potential OPSEC concerns.

The OPSEC consideration is that flying a UAS above an area the U.S. military is about to enter might give the enemy a clue about U.S. intentions. OPSEC considerations may make reliance on all-source intelligence products for planning preferable to flying UAS sorties.

Finding Supplies That Missed the Drop Zone

In this concept, cargo is being delivered by airdrop. The cargo is rigged on 463L platforms and delivered by parachute to a drop zone by aircraft from an Air Force tactical airlift squadron. In the process, a navigation error, equipment failure, unexpected winds, or other factors intervene, and one or more of the cargo platforms misses the drop zone. As a part of its search efforts to recover the missing cargo, the Army unit involved launches a UAS with SAR and FMV, which will see the large cargo parachutes or the cargo itself and thus assist the recovering unit in its task.

Case Zero

If these events occurred today, the Air Force mission commander would file a report after the mission noting that some number of platforms missed the drop zone. The Army Ground Liaison Officer (GLO) at the supporting airfield would pass the report to the Army unit in question, which would in turn mount a recovery effort. If the platforms were equipped with 3G tags and these were operational, the unit could use the resulting GPS data to locate the platforms. In the absence of such in-transit data, the unit would mount a search, probably oriented initially along the azimuth the aircraft were flying as they approached the drop zone. The unit might conduct a ground search short and long of the drop zone and on the downwind side; they might supplement the ground search with help from a RSTA or cavalry squadron, which might provide either light observation helicopters or UAS.

Once they found the cargo, the unit would begin recovery operations. Recovery might involve Pathfinders if the cargo were tangled in trees, and require helicopters for vertical extraction of the remnants under such circumstances. The recovery effort might require a security detail, depending upon the threat. Ideally, the unit would recover everything at the landing site. If, however, a significant threat of enemy action were present, the unit might have no choice but to destroy the cargo on-scene and sanitize the site to prevent enemy exploitation and intelligence production once the unit left.

Technical Feasibility

Four criteria define technical feasibility given today's UAS:

- Are there UAS capable of flying the mission?
- Are there sensors that could find the cargo?

- Can the cargo platforms somehow be instrumented to aid in locating them?
- Can the UAS carry a radio adequate to transmit cargo location information to the recovery unit?

Even the smaller tactical UAS should be capable of performing the task, depending upon the sensors they carry and the way the cargo platforms are instrumented. The point is that the U.S. military inventory of UAS is broad enough to provide a suitable aircraft.

Finding cargo platforms and large, perhaps multiple parachute canopies does not require the sensor to be capable of higher NIIRS scores; ratings of 4 or 5 should be adequate to distinguish between the cargo and parachutes and the background. FMV, perhaps in infrared, SAR would be likely candidates.

Of course, the platforms themselves could be instrumented to help the recovery unit's UAS find them. The aforementioned 3G tag might be ideal, but simple beacons (a radio signal broadcast or an IR strobe) activated upon landing could also prove useful. The parachutes might have IR tape on their canopies to make them easier to locate.

The UAS would have a radio. The recovery unit would be close by, having expected to meet the cargo at the drop zone. Communications would therefore be line-of-sight and technically very feasible given RAVEN video display units, Rover laptops, and the like.

Operational Feasibility

This concept is operationally feasible. It would require, in addition to the logistics unit's access to appropriate UAS, the establishment of standing operating procedures with the aerial delivery rigging units for marking and instrumenting cargo for ease of location once on the ground. It also depends on the availability of security forces to support recovery operations.

According to a comprehensive UAS roadmap released by the Office of the Secretary of Defense (OSD), UAS will proliferate down from division through Brigade Combat Teams (BCTs) to battalions and companies.[21] Army RSTA squadrons and similar reconnaissance formations will have organic UAS suitable for the job. Finding UAS to support recovery operations should in principle not be difficult; logistics commanders should expect, however, that their requests for UAS support will have to compete with requests from other users.

Establishing appropriate protocols with the aerial delivery community should not be difficult; after all, they are a part of the larger logistics/sustainment family.

Similarly, finding security support from maneuver formations should be operationally feasible. Infantry is regularly detailed to support convoy operations. There is no reason to believe the maneuver community would not support cargo recovery activities if the threat made doing so prudent.

---

[21] Office of the Secretary of Defense, *Unmanned Aircraft Systems Roadmap, 2005–2030,* August 4, 2005.

Sensors and Imaging

UAS can carry various sensor packages, including electro-optical and infrared sensors, SAR, SIGINT, and multi- and hyperspectral imagers. The reconnaissance and surveillance tasks envisioned in the concepts examined in this study put a premium on "real-time" sensors, or those whose results can be exploited immediately without the need for lengthy and time-consuming postprocessing and annotation in order to render them intelligible. The requirement for real-time exploitation means that UAS flying logistics reconnaissance and surveillance would be limited to electro-optical/infrared sensors, some varieties of SAR, and SIGINT. Other imaging systems exploiting different spectra cannot satisfy the real-time criterion.

Target detection, tracking, recognition, and identification based on imaging are complicated processes and are affected by many factors. One of the most significant is image resolution. Imaging sensors produce images of varying degrees of resolution. These degrees of resolution are characterized through the NIIRS as a numerical score, typically from 1 to 9, where 1 represents very-low-resolution images and 9 very-high-resolution images. Figure B.1 offers sample images with varying NIIRS scores.[1] The left-hand image, at NIIRS 5, provides enough resolution at normal size to distinguish cars and buildings, and to identify planted crops and an orchard. The center image, of a Viper Strike weapons test, approaches NIIRS 6. We can see the target car and the shadows of the towing cables between the target and the truck

---

[1] See Imagery Resolution Assessments and Reporting Standards (IRARS) Committee, "Civil NIIRS Reference Guide," March 1996.

towing it, and we can discern the roadbed from the vegetation along the berm. The image on the right, taken with a Lynx SAR in ground moving target indicating mode, falls somewhere between the images to its left. We can see buildings and cars, but much of what we perceive in the image we are able to identify through their relative positions to each other. For example, the indistinct blobs on the right middle of the image make sense to us as cars because they appear in parallel rows near squares in the image—something most viewers would interpret as cars in a parking lot adjacent to buildings.

The problem for the reconnaissance and surveillance tasks the logistics community envisions for UAS is that the available real-time sensors typically generate imagery with NIIRS ratings between 5 and 6.5.[2] Such ratings do not provide resolution high enough to discern IEDs (although there are exceptions as noted earlier in this monograph, including SAR that can sometimes sense the disturbed earth if an IED

Figure B.1
Comparative Imaging

NIIRS 5 Image

SOURCE: Federation of
American Scientists.

Hunter UAV

SOURCE: Northrop Grumman.

Lynx SAR-GMTI
0.3 Meter Imagery

SOURCE: General
Atomics.

RAND *MG978-B.1*

---

[2]   Based upon various "ISR smartbooks" we have consulted. Smartbooks are usually produced by Army and Air Force units to help planners choose the appropriate capability for the mission they are planning. A sensor's NIIRS rating could be improved by flying at lower altitudes, using longer focal lengths, and frame-averaging techniques.

is embedded in the roadway) or their command-detonation wires. At this level of resolution, it is difficult to see whether the bundle in a man's arms is an infant or an artillery shell.

As a result of this limitation on real-time sensors, UAS so instrumented today have a very difficult time seeing the enemy under circumstances that typically threaten convoys.

The future will probably see improvements in sensor technologies. Progress in miniaturization may make it possible to fly more types of sensors aboard UAS. Progress in automated imagery exploitation may reduce our dependency on human review, enabling us to take full advantage of the thousands of hours of FMV that currently remain unreviewed. As the vision articulated by the USD(I) for robust ISR bears fruit,[3] whole new constellations of sensors will appear in the skies of active military theaters.

---

[3] General Koziol's presentation at RAND, November 5, 2009.

Analysis of Cost and Benefit

According to the analysis in the previous parts of this monograph, only two concepts now enjoy technical and operational feasibility with today's UAS in conditions similar to those experienced in Iraq: finding airdropped cargo and predeployment theater reconnaissance.[1] Neither mission requires new acquisition of UAS. The logistics units can perform such UAS missions as finding airdropped cargo, but the UAS are better supplied by the owning units. So, the marginal cost to the logistics community would likely be close to zero. In addition, the operational feasibility of finding airdropped cargo may be theater-dependent. In the case of predeployment theater reconnaissance, it is operationally feasible, but the additional value of such a UAS mission would be marginal, compared with Case Zero and Case Two. There is also an OPSEC consideration. Given that there are no operational concepts that warrant cost assessment for new acquisition of UAS for logistics roles reviewed in this study, this appendix illustrates and compares the costs of a few selected concepts to show the Army the cost implications of the selected UAS mission concepts for logistics roles.

We selected two logistics convoy protection concepts for the use of UAS (armed overwatch and overhead route surveillance) and compared them with a non-UAS option: fixed surveillance. We selected these two UAS missions for the following reasons:

---

[1] Pipeline surveillance was a promising case for some combat environments but not ones that are similar to those in Iraq today.

- Reconnaissance and surveillance for force protection is one of the well-established operational uses of UAS, and readers may have expected the concepts to potentially extend to protecting logistics convoys as well.
- The sponsor organizations within the Army have indicated that these are some of the most promising areas within the UAS mission concepts assigned to this study.

After we discuss the assessment of the cost of each selected concept, we also discuss whether the benefit of such missions is likely to justify the cost. Expected outcomes or potential benefits of the two UAS missions, armed overwatch and route surveillance, are the deterrence or reduction of damage to logistics convoys and assets, which might be achieved through enhancing logistics convoys' situational awareness and augmenting armed responses against enemy attacks.

Assessing Potential Cost

We analyze a 20-year life-cycle cost for each selected concept using the present (2009) value of the dollar and the 2.9 percent discount rate specified for use by the Office of Management and Budget.[2] Cost components assessed include:

- Initial acquisition cost of the systems involved.
- Replacement costs of such systems over the life cycle that incur due to mishaps and consumption.
- Operating and support (O&S) costs.

Major O&S cost items include:

- Operational personnel: pilots, aircrew, crew technicians, and command and control personnel.
- Maintenance personnel.

---

[2]   The most recent OMB guidance (December 2008) specified using a 2.9 percent discount rate for 20-year projects.

- Contractor logistics support and sustaining engineering support.
- Indirect support: specialty training, medical support, installation support, etc.
- Fuel consumption.
- General support consumables.
- Mission support consumables.
- Repair parts, transportation, and other.

Our cost assessment draws on existing literature, publicly available information, or other sources that the study team was able to access. Different sources often provide different information for the same cost item. In such cases, we utilized all reasonable information, instead of depending on only one of those sources. This is because there is no *a priori* reason to believe that one source of information is superior to the other. For some cases, the authors depended on cost information for analogous items or turned available information into appropriate inputs for our cost assessment. Therefore, cost estimates provided in this appendix are not point estimates but ranges. The cost estimates are to help the Army have a sense of the cost of selected UAS mission concepts for logistics roles at an order of magnitude level.

Assumptions on Operational Concepts

The potential costs of each concept depend heavily on how the UAS or other alternative systems are employed and in what environments. For cost assessments, we made a set of simple assumptions about operational concepts, as well as about cost components, in addition to the assumptions presented in the main text and Appendix A.

Logistics Convoy Protection Mission

We assume the logistics convoy protection mission is performed in a similar environment to that in Iraq in the 2008 timeframe:

- Logistics convoys to be protected are moving all day long.

- Daily throughput is 24 convoys outbound, and one convoy starts every hour.
- Empty returning convoys are assumed to return on the same schedule but offset by a half hour.
- Average speed of each convoy is 40 MPH, which means that one-way spacing between convoys is 40 miles and two-way spacing is 20 miles.
- Therefore, at any given time of operation, there are approximately eight loaded convoys on the road and eight returning empty convoys.

Platforms

We assume that a Sky Warrior (MQ-1C) equipped with weapons is the most appropriate platform for the two UAS missions, although a UAS route surveillance mission can be done without armed response. This is because we intend to compare costs among different operational concepts that can achieve similar benefits. The main alternative system involved in the fixed surveillance concept consists of a combination of UH-72 light utility helicopters and rotating FLIR sensors that are mounted on poles and emplaced on the ground next to the main supply route. The non-UAS alternative (fixed surveillance) is an operational concept that can substitute for the two UAS alternatives, achieving similar benefits. Three subsections below summarize the assumptions made for each of the three operational options reviewed in this appendix.

Three Operational Options

Armed Overwatch. Each logistics convoy is escorted by a single MQ-1C Sky Warrior platform with an AAS-52 sensor package and two Hellfire missiles (AGM-114P). Given the convoy operation assumptions above, 16 MQ-1C are airborne at any time during the logistics convoy operation. Every logistics convoy is equipped with a ROVER receiver to process image data collected by the UAS. If each UAS sensor covers a ten-mile range, then approximately half of the MSR is covered by sensors at any given time. If the range of each weapon loaded to the

UAS is about ten miles, the UAS would provide near-instantaneous close air support (CAS).

Overhead Route Surveillance. The same type of MQ-1C is assigned to every 100 miles of the route with a cruise airspeed of 85 KTAS. Three MQ-1Cs are airborne at any given time of day, seven days a week and 365 days a year, over the 300-mile MSR.[3] Each logistics convoy carries a ROVER to process the data from the UAS. Assuming the ranges of the sensor and the weapon are the same as those in the armed overwatch concept above, about one-tenth of the MSR will be covered by sensors and weapons at any given time. CAS response time could be approximately 30 minutes.

Fixed Surveillance. On the 300-mile MSR, a QRF 504-person MP battalion with three line companies is separated into three fixed outpost positions at the 75th mile, 150th mile, and 225th mile. Each outpost has four UH-72 light utility helicopters for response. In addition, rotating FLIR sensors that are mounted on 100-foot poles are emplaced along the MSR at 10-mile intervals. Each logistics convoy carries a ROVER to downlink the FLIR image data. If the human target detection range of the rotating FLIR system is five miles, then this operation will provide near-continuous coverage of the whole route. CAS response time would be approximately 30 minutes.

Inputs in the Cost Analysis

This section explains the inputs used for the cost analysis for system acquisition, O&S, replacement, and munition consumption costs.

Unit Procurement Cost

Table C.1 provides the unit procurement cost of each of the systems involved in the operational concepts reviewed in this appendix, as well as cost bases and sources of information.

---

[3]   The Air Force experience tells us that we need four vehicles to keep one combat-ready vehicle in the orbit for 24/7/365 operation.

Table C.1
Unit Procurement Cost Inputs for System Procurement Cost Assessment

Cost Items	Unit Cost ($ Thousand)	Cost Basis
MQ-1C	6,500–7,625	Average procurement cost per vehicle including ground systems and satellite link. Eisman (2008), GAO (2009), U.S. Air Force (2008a).
UH-72	5,530	Average procurement cost per vehicle. Davis and Crosby (2009).
ROVER	36	Unit procurement cost. Sirack (2007).
FLIR camera	130	CCTV Imports company website.
100-foot flagpole	10	United States Flag Store website.
Munition	50	Jane's Air-Launched Weapons (2009).

Operating and Supporting Cost

There are two distinguishable types of O&S cost for UAS.[4] Some O&S cost items vary with flying hours, and the costs of other items vary with the number of systems in operation. Let us call the first type a "variable" O&S cost and the latter a "fixed" O&S cost. Examples of fixed O&S costs include: pilot, aircrew, crew technicians, maintenance, and other personnel salaries; cost of sustaining engineering support; and indirect support cost. Contractor logistics support (CLS) accounts for a large portion of UAS O&S cost, and it is primarily fixed as well.[5] Variable O&S cost items include fuel consumption, consumable material and repair parts, and depot-level reparables.

An unpublished study (Eisman, 2008) estimated the annual O&S cost of the MQ-1C Sky Warrior to be $2.35 to $2.82 million per

---

[4]   This distinction is important for our assessment as we will use existing estimates for O&S cost as inputs, adjusting flying hour differences.

[5]   More than half of the fixed O&S cost of Predator is CLS cost, according to Mel Eisman, "Air Force UAS O&S Cost Drivers, Trends, and Sustainment Challenges," briefing at RAND Logistics Seminar, August 6, 2009.

combat-ready vehicle, based on historical data for the MQ-1 Predator.[6] In a recent briefing, Eisman (2009) estimated the annual fixed O&S cost of each Predator to be approximately $3 million per vehicle and variable O&S cost to be about 35 percent of the fixed operating cost. We estimate, after additional research, that the annual O&S cost of MQ-1C Sky Warrior is within the range described below. The low-end estimate of the annual O&S cost is $2 million fixed O&S cost per combat-ready vehicle plus $1,500 per flight hour. For a high-end estimate, the fixed O&S cost is assumed to be $3 million per combat-ready vehicle plus $2,000 per flight hour. We estimate the UAS system O&S cost with sensors.

For the UH-72 O&S cost, we estimate approximately $900,000 in personnel per vehicle per year and $2,200 per flight hour, based on calculations employing information from various public sources.[7] In examining FLIR sensors for the fixed surveillance concept, we assume their annual O&S cost plus replacement cost to be 30–35 percent of the procurement cost. ROVER annual O&S cost plus replacement cost is assumed to be 25–30 percent of the procurement cost. The annual cost of leasing a commercial satellite communication network is assumed to be $500,000 per orbit.[8]

Mishap Rates and Replacement Cost

For UAS, we assume each vehicle needs to be replaced due to losses every 10,000–11,000 flying hours, based on the historical data from Class A mishaps for Predator and from data on the vehicle's cumulative flying hours. There had been 37 Class A mishaps for Predator vehicles from fiscal year (FY) 2000 through FY 2008, according to the Accident Investigation Board (AIB) reports by the U.S. Air Force

---

[6] Historical data on Predator are often used to estimate Sky Warrior costs because Sky Warrior is a system analogous to Predator.

[7] Sources used include: Conklin and de Decker Aviation Information's website as of September 1, 2009, and Michael L. Wesolek, "Army Aviators Better Trained, but at Higher Costs," *National Defense*, June 2006.

[8] Currently, the Predator uses commercial satellite communication (SATCOM) networks. According to the Air Force's UAV Flight Plan 2009–2047 (2009), even by 2018, 54 percent of UAS usage will still rely on leases of civilian SATCOM.

Judge Advocate General's Corps.[9] As of October 2008, there were thirty-one total Combat Air Patrols (CAPs) (which includes six Air Force Special Operations Command and eight Air National Guard) of MQ-1 Predator vehicles deployed with a cumulative total of 400,000 of combat flying hours.[10] This means there was one Predator mishap for every 10,810 flying hours of combat missions on average. Our mishap assumption for Sky Warrior is within the range of the Predator experience but slightly more optimistic.

For the UH-72, we assume that the replacement of aircraft happens every 54,000 flying hours, using a similar reasoning to the UAS case above. The UH-72 is not a mature system, not yet having flown 100,000 operational hours. Therefore, we used the average Class A mishap rate for all Army helicopters as of September 13, 2009. On average, there were 1.858 Class A flying incidents per 100,000 flight hours in the last three years.[11] Frequency of munition consumption is assumed to be once every week or so based on our interviews with Army logisticians who have recently come back from Iraq.

Because the FLIR sensors would be mounted on poles, and the poles would be subject to wear and tear, we assumed their annual replacement cost is approximately 20 percent of the procurement costs. This assumption is based on the typical depreciation rule in accounting that assumes purchased goods are depreciated within five years.

Cost Analysis Results

Based on the inputs above, we estimated the 20-year life-cycle costs in present value terms for the three operational options. The results are summarized in Table C.2.

---

[9] The reports are available at the USAF Judge Advocate General's Corps website: "United States Air Force Class A Aerospace Mishaps."

[10] Briefing by the Air Force's ACC/A8M Capabilities Division, October 2008.

[11] U.S. Army Accident Information "Army Aviation Accident Year-end Data," September 13, 2009.

Table C.2
Cost Comparison Among the Three Operational Alternatives for Logistics
Convoy Protection

	Armed Overwatch	Overhead Route Surveillance	Fixed Surveillance
Life-cycle cost in present value	2,385–3,314	773–1,058	667–718

NOTES: Base year is 2009. Discount rate is 2.9 percent. Unit: U.S. $ million.

Table C.2 shows that the overhead route surveillance by UAS is estimated to be 32 percent more expensive than the fixed surveillance alternative. As expected, the armed overwatch is much more expensive than other options. Compared to the route surveillance by UAS, the life-cycle cost of the armed overwatch alternative is approximately three times more expensive. We also roughly estimated the cost of leasing for each of the operational alternatives and concluded that leasing would be more expensive than owning the systems.

Cost Estimates with Leasing Arrangements

Instead of purchasing, the Army may want to consider leasing UAS for specific logistics applications. However, price quotes were not available for leasing UAS operations. Leasing price information for a medium-altitude and medium-endurance UAS that is close to Sky Warrior can be found in the case of the Canadian Air Force's lease of the Heron UAS. In 2008, Israel Aerospace Industries (IAI) and MacDonald Dettwiller and Association Ltd. (MDA) were awarded a leasing contract from the Canadian Air Force to deliver the CU-170 Heron UAS to Canadian forces deployed in Afghanistan.[12] The two-year lease price of the UAS per orbit was 81 million U.S. dollars (95 million Canadian

---

[12] Refer to "Israel Aerospace Industries announces delivery of the 1st Heron UAV to Canada," *Shephard: News*, web page, October 15, 2008; National Defence and the Canadian Forces, "Newly-Acquired UAV Headed for Afghanistan," December 18, 2008; and "CF Leased UAV—McDonald Dettwiler/IAI Malat CU-170 Heron," *Canadian American Strategic Review*, March 12, 2009.

dollars). The lease price included the contractors' cost of management, training, and in-theater maintenance for the Canadian operators. The operators were to pay for operating crews (air vehicle operators, payload operators, intelligence analysts) and fuel cost. Based on the information available from this case, we estimated the costs of two UAS alternatives (armed overwatch and overhead surveillance) with leasing arrangements similar to the Canadian case. The results are summarized in Table C.3.

Table C.3
Cost Estimates for UAS Alternatives with Leasing Arrangements

	Armed Overwatch	Overhead Route Surveillance
Annual cost	729–748	139–145
Life-cycle cost	10,950–11,236	2,080–2,173

NOTES: Base year is 2009. Discount rate is 2.9 percent. For a lease longer than two years, we assumed the Army gets a discount in each year's leasing price up to the annual inflation rate. Unit: U.S. $ million.

The results above show that in terms of life-cycle cost, UAS leasing arrangements are probably much more expensive than purchasing UAS. Leasing arrangements could be attractive if the Army used them only for a short period. Even if the lease arrangements are made for a short-term usage, the leased system can produce all sorts of DOT-MLPF issues when inserted into a military formation.

Comparing Potential Costs with Potential Benefits

Benefit analysis is useful for several reasons. First, it can help the Army determine whether the benefits of the selected operational concepts are likely to justify the costs. Second, instead of acquiring new UAS systems that are dedicated to logistics missions, the COCOM may want to use existing UAS systems. In this case, the Army needs to know whether the potential benefit of using UAS for logistics convoy protection is significant enough to replace one of the current nonlogistics UAS missions.

Assessing Potential Benefits

Potential benefits of logistics convoy protection depend on three main variables:

- Frequency of attacks on logistics convoys.
- Expected average damage of each attack.
- Effectiveness of UAS in either preventing such attacks or reducing the expected damages of attacks.

Given that we do not have clear foresight of the future, let us try to understand the frequency of attacks and related damages, assuming the future combat environments would be similar to the ones in Iraq and Afghanistan. We found a set of classified data on improvised explosive devices (IEDs) and other anti-armor attacks that helped us understand the frequency of such attacks on logistics convoys, and the extent of damages due to such attacks during the last several years. The data include types of attacks, types of missions attacked, number and types of vehicles damaged, the extent of damage to each vehicle, and number of soldiers wounded or killed in action because of the attacks. By using these data and estimating the damage in dollar terms, we calculated estimates for expected annual order-of-magnitude damages to logistics convoys.

The next question is to determine how effective the UAS will be in deterring or reducing the damages to logistics convoys. The effectiveness of UAS in such missions, of course, would depend on what types of UAS are employed as well as "how," "where," "when," and "why." To understand UAS mission effectiveness, we searched for existing quantitative or qualitative data that describe the contributions of UAS to surveillance and force protection missions. However, we were not able to find such data. Therefore, we take a parametric approach to examine UAS effectiveness.

Informing Breakeven Points

Comparing potential costs with potential benefits can provide information about whether the UAS alternatives are cost-effective. Given the potential cost estimates for each operational concept, and the estimates of expected damages to logistics convoys, the breakeven points depend on UAS mission effectiveness and on the probability that a convoy is attacked.

Let us set P_a as the probability that a convoy is attacked, D_a as the average dollar value damage resulting from each attack, and f as an indicator of mission effectiveness. Then,

$P_a D_a$ = Expected damage per convoy of attack in Case Zero.

$(1 - f)\, P_a D_a$ = Expected damage per convoy of attack in Case One with UAS armed overwatch or route surveillance, where

$$0 < f \leq 1.$$

The incremental benefit *per convoy* of having armed overwatch or route surveillance by a UAS is

$$P_a D_a f.$$

The incremental benefit over the time horizon of the UAS project is

$$PV(\text{Benefit}) = PV\, \{P_{at} \cdot D_{at} \cdot f\} \cdot N_t\}_{t=1}^{n},$$

where

PV   = Present value.

N_t = Number of convoys in year t.

n   = Time horizon (number of years).

P_{at} = Probability that a convoy is attacked in year t.

D_{at} = Damage of attack in dollars in year t.

Then, breakeven points can be obtained from the following:

$$PV \text{ (Benefit)} = PV(\text{cost of the UAS project}) = K,$$

where K is the estimated cost in present value in 2009 dollars.

To illustrate a simple case, let us assume:

N_t is identical for all years, as N_0.

P_{at} is identical for all years, as P_a.

D_{at} is identical for all years, as D_a.

Then, the annual average benefit of having overwatch by a UAS is $P_a D_a f N_0$. For each operational concept to be cost-effective, the annual average benefit should be larger than or equal to an annualized potential cost,

$$K_1 = K \sum_{t=1}^{n} (1 + r)^t.$$

If we have a good sense of what D_a and N_0 would be, then we can derive the following:

$$P_a \cdot f = K_2,$$

where

$$K_2 = K_1 / D_a N_0.$$

With this formula, we can draw indifference curves of breakeven points where the horizontal and vertical axes are f and P_a. The graph then helps one visualize whether and under what ranges of risk of attack (P_a) and UAS mission effectiveness (f) the armed overwatch and route surveillance missions of UAS would be likely to make economic sense. Relatively high levels of risk of attack and UAS mission effectiveness are required for the selected UAS operational concepts to be economically justified.

Unmanned Aircraft Systems Overview

This appendix provides a short overview of UAS attributes. Readers looking for a more detailed description should consult the OSD UAS Roadmap.[1]

Table D.1 summarizes characteristics of typical UAS. It reflects fixed-wing and rotorcraft, Army organic systems, and other systems likely to be encountered in theater.

Table D.1
Common UAS

Name/ Designation	Max Time Aloft	Range (radius)	Payload Weight
Raven RQ-11B	90 minutes	10 km	6.5 oz
Scan Eagle	23 hours	100 km	13 lbs
Shadow RQ-7B	7 hours	110 km	100 lbs
I-Gnat ER/Sky Warrior Alpha	40 hours	5,400 km with comms relay	800 lbs
ERMP MQ-1C Sky Warrior Block 0	30 hours	3,750 km with comms relay	800 lbs
Rotorcraft:			
A-160 Hummingbird	24 hours	4,023 km	1,000 lbs
Fire Scout	8 hours	203 km	500 lbs
Steadycopter Black Eagle 50	4 hours	260 km	22 lbs
Alternative small UAS:			
Silver Fox	12 hours	32 km	4 lbs
Killer Bee	30 hours	100 km	20 lbs

---

[1] Office of the Secretary of Defense, August 4, 2005.

The Raven, Shadow, and Sky Warriors are found in Army units. The rotorcraft and alternative UAS listed below them are not, although they are in Marine Corps and Navy organizations and therefore often available in Iraq and Afghanistan. Readers who need mission planning quality information on UAS should consult the many Army and Air Force "smartbooks" available via SIPRNET.

Alternatives for Tasks Where UAS Concepts Appear Infeasible

The analysis found a number of the UAS concepts operationally infeasible and thus eliminated the need to examine potential benefits and costs. The question at this point is what other remedies might assist the logistics community today with the various tasks under discussion, i.e., protecting convoys, securing pipelines, ascertaining the navigability of rivers, securing logistics sites, and supporting operations following a mass casualty domestic disaster. This appendix offers some preliminary thoughts on the answer, while acknowledging that the bulk of the analytical effort has been to examine the UAS concepts rather than produce exhaustive alternatives to them.

Protecting Convoys

According to Peters, Bonds, and Fischbach (2010), U.S. forces in Iraq found about 40 percent of the IEDs the enemy deployed between 2004 and 2007.[1] According to the Joint Improvised Explosive Device Defeat Organization (JIEDDO) 2008 annual report, the percentage discovered before detonation is slightly higher in Afghanistan, around 50 percent.[2] In both cases, however, the numbers deployed fluctuate sig-

---

[1] Derived from CIDNE and FusionNet data.

[2] Joint Improvised Explosive Device Defeat Organization (JIEDDO), *Annual Report FY 2008*, Figure 3.

nificantly from month to month. The percentages of IEDs discovered do not vary much, typically plus-or-minus 5 or 7 percent, suggesting there is a limit to the efficiency at which current counter-IED technology can detect these weapons. Thus, if logistics convoys require additional protection beyond what regular route clearance, route surveillance, armor, and TTPs can provide, the options are limited but not inconsequential.

First, the logistics community can take greater advantage of tip lines, where the local populace can call joint coordination centers to report suspicious activities and warn of IEDs.[3] Coordination with military police, psychological operations units, intelligence units, and indigenous forces can produce unity of action and mobilize the population to support freedom of movement along the highways.

Second, it may become necessary to place greater emphasis on control of the highways by U.S. or friendly forces. Positive control of lines of communication has long been a crucial part of counterinsurgency operations and has often required fortifications along the route to maintain security. Pakistan has taken steps toward fortifying the approaches to the Khyber Pass for this reason, and the British built blockhouses at intervisible distances to protect the railroad during their mandate over what is today Iraq during the 1920 insurrection.[4] The United States has done the same thing with the route between the Green Zone and Baghdad International Airport.

Third, the logistics community could avail itself of some of the developments to emerge from JIEDDO. Two in particular stand out: Palantir and Vehicle Optics Sensor System (VOSS). Palantir is a network and collaborative network analysis tool "used to identify patterns and relationships between entities and events (for counter-IED

---

[3] See, for example, Major Gordon J. Knowles, "Countering Terrorist, Insurgent, and Criminal Organizations: Iraqi Security Forces Joint Coordination Centers—A Unique Public Safety System," *Military Police*, PB 19-06-2, Fall 2006.

[4] Gertrude Bell, *The Arab War: Confidential Information for General Headquarters from Gertrude Bell Being Dispatches Reprinted from the Secret "Arab Bulletin,"* London: Golden Cockerel Press, 1940.

purposes)."[5] VOSS features a mast-mounted, stabilized camera on a vehicle for day and night operations. The system allows the operator to "scan for IED indicators while on the move . . . and to interrogate suspected targets from a safe standoff range . . . ."[6] Palantir would equip logistics units to fully exploit intelligence about IEDs on their LOCs, and VOSS would help convoys avoid roadside threats.

Another option might be to launch an offensive counter-IED campaign. TF ODIN and Project Liberty have proved successful in this regard, actively hunting for IED emplacement teams and catching them in the act. They employ manned/unmanned teaming and may have killed as many as 3,000 enemy combatants.

Finally, if a route cannot be made sufficiently secure to accommodate logistics convoys, then it should be abandoned and an alternate route chosen that can be secured. If ground transportation remains too dangerous, then it may be necessary to move as much traffic to the air as possible, although such a move would clearly have consequences for the overall conduct of the campaign.

Protecting Pipelines

Attacks require very limited dwell time, and the perpetrators do not routinely appear with weapons, which limits the ability of UAS-borne sensors to identify hostile intent in the rare instances in which the UAS is actually overhead while the attackers prepare to do their work. A more effective security architecture would involve instrumenting the pipeline to alert authorities of tampering. The Army currently has a number of systems that provide photographic, magnetic, acoustic, and seismic data. Scorpion, Silent Watch, Omnisensor, Vistas, and the Tactical Remote Sensor System represent the newest suites of capabilities; the Army also has the long-standing Remote Battlefield Sensing System (REMBASS) still in its inventory. Pipelines thus instrumented could provide initial alerts, and the responsible Army unit could fly a

---

[5]   JIEDDO Annual Report, FY 2008, p. 9.

[6]   JIEDDO Annual Report, FY 2008, p. 11.

UAS out to the scene to follow the enemy as he makes his escape, passing location data to a ground QRF that could follow and detain the attackers.

According to U.S. contractors operating in Iraq, tribal militias can in some instances provide credible security. Based on contractor experience, some groups are more reliable than others, but even mediocre militias can be cajoled into maintaining power line security.[7] Apparently, tribal connections and loyalties support power line security from a shared-interests point of view—i.e., if it is important to the sheik, then it is important to us. Tribal dynamics also seem to make intimidation possible, so that actors who might otherwise attack the power line will, upon learning that it is under the protection of a given militia or tribe, look elsewhere for sabotage targets.

Ascertaining Navigability of Rivers

Determining the navigability of a river with a UAS-borne sensor was the only concept that failed the technical feasibility test. There are, however, established alternatives. As a part of its deliberate planning, the logistics community could generate an intelligence-collection requirement to determine the navigability of rivers of interest. On a more immediate-need basis, the Army could ask the U.S. Coast Guard and Navy for their assistance, given that both services operate autonomous survey systems. Finally, the logistics community could take advantage of extant operational support documents and GIS data to determine the navigability of a river of interest.

Protecting Logistics Facilities

The best options for protecting fixed installations, whether they are depots, LSAs, bladder farms, maintenance facilities, or other logistics

---

[7] Interview with officials from Arctic Slope Airfield and Range Services (ASARS) operating in Baghdad and Taji, Iraq, January 20, 2009.

facilities, involve developing an integrated security system for the site that exploits the capabilities of various radars, cameras, and sensors to provide situational awareness and to detect incoming rocket, artillery, and mortar fire, and unauthorized approaches to the perimeter (e.g., infiltrators, VBIEDs).[8] Camera towers and aerostat-borne sensors typically provide higher levels of resolution than real-time UAS sensors and are easier to maintain than UAS orbits. Surveillance data must be supplemented by other intelligence, especially HUMINT; for best results, feedback among and between the intelligence disciplines deployed is critical. The tactical automated security system (TASS) is essential to provide sufficient warning time for personnel near the predicted point of impact of an indirect fire attack to take cover. Terminal defenses from the Army's C-RAM program offer some hope of destroying incoming ordnance before it detonates. Such a system—intelligence and surveillance, active and passive defenses, and a post-attack response—works best when all the data feeds and images are integrated in a base defense operations center. The best role for UAS in such a defensive system appears to be in tracking the assailants in the aftermath of the attack and leading the QRF to them, although the data sample underpinning this conclusion is small.

---

[8]   Based upon analysis in Chow et al. (2009) and Peters et al. (2010).

Bibliography

110th Congress 1st Session1351, *National Defense Authorization Act for Fiscal Year 2008*, 35-737, Senate Report 110-77, June 5, 2007.

"Army Praises ODIN," *Aviation Week*, May 14, 2007.

Association for Unmanned Vehicle Systems International, website. As of June 2010:
http://www.auvsi.org/AUVSI/AUVSI/Home

Bell, Gertrude, *The Arab War: Confidential Information for General Headquarters from Gertrude Bell Being Dispatches Reprinted from the Secret "Arab Bulletin,"* London: Golden Cockerel Press, 1940.

Bonds, Timothy M. John E. Peters, Endy Y. Min, Lionel A. Galway, Jordan R. Fischbach, Eric Gons, Garrett D. Heath, and Jean M. Jones, "Army Network-Enabled Operations: Expectations, Performance, and Opportunities for Future Improvements," unpublished RAND Corporation research, 2010.

Brady, Matthew D., "The 766th Explosive Hazards Coordination Cell Leads the Way into Afghanistan," *Engineer*, January–April 2009, pp. 42–47.

Brock, Lt Col Bob, USAF, "3d SOS Command Brief: UAS Vision and Frameworks," April 20, 2009.

CCTV Imports "Ranger III XR+ Long Range Thermal Security Camera—150-750mm," CCTV Imports company website, undated. As of September 12, 2009:
http://www.cctvimports.com/cameras-flir-thermal-security-cameras-c-3_252/ranger-iii-xr-long-range-thermal-security-camera-150-750mm-p-966

"CF Leased UAV—McDonald Dettwiler/IAI Malat CU-170 Heron," *Canadian American Strategic Review*, March 12, 2009. As of September 12, 2009:
http://www.casr.ca/101-af-cu170-heron-uav.htm

Chow, Brian G., John E. Peters, Katherine Comanor, Marvin Schaffer, and Edward R. Harshberger, *Fighting Air Bases Under Attack: Forward Operating Bases (U)*, Santa Monica, Calif.: RAND Corporation, 2009. Not available to the general public.

Congress of the United States, Congressional Budget Office, "The Army's Bandwidth Bottleneck," August 2003.

Conklin and de Decker Aviation Information, "Helicopter Variable Cost Data," undated. As of September 1, 2009:
http://www.conklindd.com/Page.aspx?cid=1118

Davis, Walter L. (Director of Army Aviation, Office of the Deputy Chief of Staff G-3/5/7), and William T. Grosby (Program Executive Officer, Aviation), "Statements," before the Air and Land forces Subcommittee, Committee on Armed Services, United States House of Representatives, Army Aviation Programs.

Department of Defense, *Joint Concept of Operations for Unmanned Aircraft Systems*, Washington, D.C.: Joint Unmanned Aircraft Systems Center of Excellence, November 2008.

Diener, David, Eric Peltz, Arthur Lackey, Darlene J. Blake, and Karthik Vaidyanathan, *Value Recovery from the Reverse Logistics Pipeline*, Santa Monica, Calif.: RAND Corporation, MG-238-A, 2004.
http://www.rand.org/pubs/monographs/MG238/index.html

Eisman, Mel, unpublished research on fighting air bases under attack, 2008.

Eisman, Mel, "Air Force UAS O&S Cost Drivers, Trends, and Sustainment Challenges," briefing at RAND Logistics Seminar, August 6, 2009.

General Dynamics, *AR 5-5 Study: Future Modular Force Resupply Mission for Unmanned Aircraft Systems (UAS)*, prepared for Commanding General, Combined Arms Support Command and Department of the Army, G-4, General Dynamics Information Technology, February 24, 2010.

Gorman, Siobhan, Yochi J. Dreazen, and August Cole, "Insurgents Hack U.S. Drones," *Wall Street Journal*, December 17, 2009.

Government Accountability Office (GAO), *Defense Acquisitions: Opportunities Exist to Achieve Greater Commonality and Efficiencies Among Unmanned Aircraft Systems*, Report to the Subcommittee on Air and Land Forces, Committee on Armed Services, House of Representatives, GAO-09-520, July 2009.

Headquarters, Department of the Army, *Mechanized Infantry Platoon and Squad (Bradley)*, Washington, D.C., Field Manual 3-21.71, August 20, 2002.

Headquarters, USA CASCOM, *Logistics Re-supply Mission Support Role for the Unmanned Aircraft System (UAS) Requirements Identification and Definition*, Information Paper, January 31, 2008.

Health Systems Research, Inc., *Altered Standards of Care in Mass Casualty Events*, Rockville, Md.: Agency for Healthcare Research and Quality, U.S. Department of Health and Human Services, AHRQ Publication No. 05-0043, April 2005. As of May 11, 2010:
http://www.ahrq.gov/research/altstand/

Imagery Resolution Assessments and Reporting Standards (IRARS) Committee, "Civil NIIRS Reference Guide," March 1996. As of May 11, 2010:
http://www.fas.org/irp/imint/niirs_c/guide.htm

"Israel Aerospace Industries Announces Delivery of the 1st Heron UAV to Canada," *Shephard: News*, web page, October 15, 2008. As of September 12, 2009:
http://www.shephard.co.uk/news/390/
israel-aerospace-industries-announces-delivery-of-the-1st-heron-uav-to-canada/

Jane's Air-Launched Weapons, "GBU-39/B Small Diameter Bomb (SDB) and SDB II (United States)," June 5, 2009. As of September 11, 2009:
http://www.janes.com/articles/Janes-Air-Launched-Weapons/GBU-39-B-Small-Diameter-Bomb-SDB-and-SDB-II-United-States.html [subscription required]

Jane's Electro-Optic Systems, "Northrop Grumman AN/PED-1 Lightweight Laser Designator Rangefinder (LLDR)," October 13, 2008. As of September 11, 2009:
http://www.janes.com/articles/Janes-Electro-Optic-Systems/Northrop-Grumman-AN-PED-1-Lightweight-Laser-Designator-Rangefinder-LLDR-United-States.html [subscription required]

Jane's Electronic Mission Aircraft, "Lynx," June 18, 2010. As of July 2010:
http://search.janes.com/Search/documentView.do?docId=/content1/janesdata/binder/jema/jemaa132.htm@current&pageSelected=allJanes&keyword=lynx%20radar&backPath=http://search.janes.com/Search&Prod_Name=JEMA&

Joe, Leland, and Isaac R. Porche, III, *Future Army Bandwidth Needs and Capabilities*, Santa Monica, Calif.: RAND Corporation, MG-156-A, 2004.
http://www.rand.org/pubs/monographs/MG156/index.html

Joint Improvised Explosive Device Defeat Organization (JIEDDO), *Annual Report FY 2008*. As of June 2010:
http://www.jieddo.dod.mil

Joint Chiefs of Staff, *Joint Targeting*, Joint Pub 3-60, April 13, 2007. As of July 2010:
http://www.dtic.mil/doctrine/new_pubs/jp3_60.pdf

Knowles, Major Gordon J., "Countering Terrorist, Insurgent, and Criminal Organizations: Iraqi Security Forces Joint Coordination Centers—A Unique Public Safety System," *Military Police*, PB 19-06-2, Fall 2006. As of July 2010:
http://www.wood.army.mil/mpbulletin/pdfs/fall%2006/Knowles.pdf

Larson, Eric V., and John E. Peters, *Preparing the U.S. Army for Homeland Security: Concepts, Issues, and Options*, Santa Monica, Calif.: RAND Corporation, MR-1251-A, 2001.
http://www.rand.org/pubs/monograph_reports/MR1251/index.html

National Defence and the Canadian Forces, "Newly-Acquired UAV Headed for Afghanistan," December 18, 2008. As of September 12, 2009:
http://www.army.forces.gc.ca/land-terre/news-nouvelles/story-reportage-eng.asp?id=3076

Oaks, David M., Matthew Stafford, and Bradley Wilson, *The Value and Impacts of Alternative Fuel Distribution Concepts: Assessing the Army's Future Needs for Temporary Fuel Pipelines*, Santa Monica, Calif.: RAND Corporation, TR-652-A, 2009.
http://www.rand.org/pubs/technical_reports/TR652/index.html

Office of Management and Budget, *Discount Rates for Cost-Effectiveness, Lease Purchase and Related Analyses*, Washington, D.C.: Office of Management and Budget, Advisory Circular A-94 Appendix C, December 2008.

Office of the Secretary of Defense, *Unmanned Aircraft Systems Roadmap, 2005–2030*, August 4, 2005. As of May 2010:
http://www.fas.org/irp/program/collect/uav_roadmap2005.pdf

Osborn, Kris, "Pentagon Readies 3 Anti-IED UAVs," *Navy Times*, December 16, 2008. As of May 11, 2010:
http://www.navytimes.com/news/2008/12/defense_IED_UAVs_121608/

Pagels, Dr. Michael A., DARPA, "Heterogeneous Airborne Reconnaissance Team (HART)," briefing, dated August 2008.

Peters, John E., et al., unpublished research for the Army on estimating the life-cycle cost of the multipurpose MQ-1C system, 2009. Not available to the general public.

Peters, John E., Timothy M. Bonds, and Jordan R. Fischbach, "Army Network Performance in Iraq: What Data Mining Suggests (U)," unpublished RAND Corporation research, 2010. Not available to the general public.

Peters, John E., Seng Boey, Harun Dogo, Diana Dunham-Scott, Daniel Gonzales, Thomas Hamilton, Jody Jacobs, Nicholas C. Maynard, Endy Y. Min, Louis R. Moore, and Thomas Sullivan, "Overhead Reconnaissance, Intelligence, and Target Acquisition Support to Army Distributed Counterinsurgency Efforts," unpublished RAND Corporation research, 2010. Not available to the general public.

Quinlivan, James, "Force Requirements in Stability Operations," *Parameters*, Winter 1995, pp. 59–69.

Scott, Richard, "DARPA Takes Advantage of TAILWIND for UAV Surveillance," *International Defense Review*, May 20, 2009.

Shachtman, Noah, "Air Force to Unleash 'Gorgon Stare' on Squirting Targets." As of February 19, 2009:
http://www.wired.com/dangerroom/2009/02/gorgon.stare/

Sirak, Michael, "Air Force Eyes Handheld Rover Wireless Video Receiver," *Defense Daily*, July 16, 2007. As of September 2, 2009:
http://findarticles.com/p/articles/mi_6712/is_10_235/ai_n29367924/

"Sky-Warrior ERMP UAV System," *Defense Update*, fact sheet. As of May 11, 2010:
http://defense-update.com/products/w/warriorUAV.htm

Trimble, Stephen, "USAF to Unleash 'Gorgon Stare' Sensor in 2010," *Flight International,* January 28, 2009. As of June 2010:
http://www.flightglobal.com/articles/2009/01/28/321732/usaf-to-unleash-gorgon-stare-sensor-in-2010.html

Trimble, Stephen, *TRANSCRIPT: Hunter Green Dart Q&A*, transcript of Tim Owings, Army Deputy Program Manager for UAVs, February 6, 2009. As of May 11, 2010:
http://www.flightglobal.com/blogs/the-dewline/2009/02/transcript-hunter-green-dart-q.html

U.S. Air Force, "Factsheets: MQ-1 Predator Unmanned Aircraft System," September 2008a. As of September 11, 2009:
http://www.af.mil/information/factsheets/factsheet.asp?id=122

U.S. Air Force, "Factsheets: MQ-9 Reaper Unmanned Aircraft System," September 2008b. As of September 11, 2009:
http://www.af.mil/information/factsheets/factsheet.asp?id=6405

U.S. Air Force, "Unmanned Aircraft Systems Flight Plan 2009–2047," May 18, 2009. As of September 11, 2009:
http://www.globalsecurity.org/military/library/policy/usaf/usaf-uas-flight-plan_2009-2047.pdf

U.S. Air Force, Air Combat Control, "Unmanned Aerial System MQ-1 Predator," Battlespace Awareness ISR Annex to the Combat Air Forces Strategic Plan, Briefing on the Air Force Portal at World Wide Web, October 2008. As of September 4, 2009 (login required):
https://wwwd.my.af.mil/afknprod/ASPs/DocMan/DOCMain.asp?Tab=0&FolderID=OO-XP-AC-65-3-3&Filter=OO-XP-AC-65

U.S. Air Force, Judge Advocate General's Corps, "United States Air Force Class A Aerospace Mishaps," Accident Investigation Board Reports, Fiscal Year 2000 through Fiscal Year 2009. As of September 3, 2009:
http://usaf.aib.law.af.mil/

U.S. Army Accident Information, "Army Aviation Accident Year-end Data," September 13, 2009. As of September 13, 2009:
https://rmis.army.mil/stats/prc_fy_aviation_stats

U.S. Army, Table of Organization and Equipment, Light Utility Aviation Company, Light Utility Aviation Battalion, Corps Aviation Group, January 1997. As of September 4, 2009:
http://www.fas.org/man/dod-101/army/unit/toe/01457A000.htm

U.S. Army Training and Doctrine Command, U.S. Army UAS Center of Excellence, *"Eyes of the Army": U.S. Army Roadmap for Unmanned Aircraft Systems,*

2010–2035, no date. As of May 2010:
http://www.rucker.army.mil/usaace/uas/US%20Army%20UAS%20RoadMap%20
2010%202035.pdf

United States Flag Store, website. As of September 3, 2009:
http://www.united-states-flag.com/grsetdisest1.html?productid=grsetdisest1&chan
nelid=FROOG

Wang, Mark Y. D., Carol E. Fan, Darlene J. Blake, Arthur M. Bullock, and Eric
Peltz, "Improving the Army's Retrograde Distribution Management Operations:
Developing a Future Vision," unpublished RAND Corporation research, 2006.

Warwick, Graham, "Staying Up, Staring Down, LEMV Airship," in "Ares—A
Defense Technology Blog," *Aviation Week,* June 8, 2009. As of May 7, 2010:
http://www.aviationweek.com/aw/blogs/defense/index.jsp?plckController=Blog&p
lckScript=blogScript&plckElementId=blogDest&plckBlogPage=BlogViewPost&p
lckPostId=Blog%3A27ec4a53-dcc8-42d0-bd3a-01329aef79a7Post%3A3c0ec60a-
4bb5-43d4-b0ff-fefdf0733cb6

Wesolek, Michael L., "Army Aviators Better Trained, but at Higher Costs,"
National Defense, June 2006. As of September 3, 2009:
http://www.nationaldefensemagazine.org/ARCHIVE/2006/JUNE/Pages/
ArmyAviaters2950.aspx

"What's New at Brimrose! SMART PROCurement 2009 Trade Show," *Brimrose,*
web page. As of May 11, 2010:
http://www.brimrose.com/whats_new.html